McGraw-Hill's

500

College Algebra

and

Trigonometry

Questions

D1530600

Also in McGraw-Hill's 500 Questions Series

McGraw-Hill's 500 American Government Questions: Ace Your College Exams
McGraw-Hill's 500 Calculus Questions: Ace Your College Exams
McGraw-Hill's 500 College Algebra and Trigonometry Questions: Ace Your College Exams
McGraw-Hill's 500 College Biology Questions: Ace Your College Exams
McGraw-Hill's 500 College Chemistry Questions: Ace Your College Exams
McGraw-Hill's 500 College Physics Questions: Ace Your College Exams
McGraw-Hill's 500 Differential Equations Questions: Ace Your College Exams
McGraw-Hill's 500 European History Questions: Ace Your College Exams
McGraw-Hill's 500 French Questions: Ace Your College Exams
McGraw-Hill's 500 Linear Algebra Questions: Ace Your College Exams
McGraw-Hill's 500 Macroeconomics Questions: Ace Your College Exams
McGraw-Hill's 500 Microeconomics Questions: Ace Your College Exams
McGraw-Hill's 500 Organic Chemistry Questions: Ace Your College Exams
McGraw-Hill's 500 Philosophy Questions: Ace Your College Exams
McGraw-Hill's 500 Physical Chemistry Questions: Ace Your College Exams
McGraw-Hill's 500 Precalculus Questions: Ace Your College Exams
McGraw-Hill's 500 Psychology Questions: Ace Your College Exams
McGraw-Hill's 500 Spanish Questions: Ace Your College Exams
McGraw-Hill's 500 U.S. History Questions, Volume 1: Ace Your College Exams
McGraw-Hill's 500 U.S. History Questions, Volume 2: Ace Your College Exams
McGraw-Hill's 500 World History Questions, Volume 1: Ace Your College Exams
McGraw-Hill's 500 World History Questions, Volume 2: Ace Your College Exams
McGraw-Hill's 500 MCAT Biology Questions to Know by Test Day
McGraw-Hill's 500 MCAT General Chemistry Questions to Know by Test Day
McGraw-Hill's 500 MCAT Physics Questions to Know by Test Day

McGraw-Hill's

500

College Algebra
and
Trigonometry
Questions

Ace Your College Exams

Philip Schmidt, PhD

New York   Chicago   San Francisco   Athens   London   Madrid
Mexico City   Milan   New Delhi   Singapore   Sydney   Toronto

Philip Schmidt has a BS from Brooklyn College and an MA in mathematics and a PhD in mathematics education from Syracuse University. He is currently professor of secondary education at SUNY New Paltz, where he coordinates the secondary mathematics education program. He is the author of *3000 Solved Problems in Precalculus* as well as numerous journal articles.

Edited by Dr. Sandra Luna McCune.

1 2 3 4 5 6 7 8 9 10 11 12 13 14 15   QFR/QFR   1 0 9 8 7 6 5 4 3

ISBN   978-0-07-178955-4
MHID   0-07-178955-3

e-ISBN   978-0-07-178956-1
e-MHID   0-07-178956-1

Library of Congress Control Number 2012933639

McGraw-Hill Education products are available at special quantity discounts to use as premiums and sales promotions or for use in corporate training programs. To contact a representative, please e-mail us at bulksales@mcgraw-hill.com.

This book is printed on acid-free paper.

CONTENTS

Introduction   vii

Chapter 1 Review of Basic Algebra   1
Questions 1–40

Chapter 2 Equations and Inequalities   5
Questions 41–101

Chapter 3 Graphs, Relations, and Functions   11
Questions 102–151

Chapter 4 Polynomial and Rational Functions   21
Questions 152–196

Chapter 5 Systems of Equations and Inequalities   27
Questions 197–246

Chapter 6 Exponential and Logarithmic Functions   33
Questions 247–276

Chapter 7 Trigonometric Functions   37
Questions 277–321

Chapter 8 Trigonometric Equations and Inequalities   41
Questions 322–366

Chapter 9 Additional Topics in Trigonometry   47
Questions 367–401

Chapter 10 Conic Sections   53
Questions 402–436

Chapter 11 The Complex Numbers   57
Questions 437–456

Chapter 12 Sequences, Series, and Probability   59
Questions 457–500

Answers   65

INTRODUCTION

Congratulations! You've taken a big step toward achieving your best grade by purchasing *McGraw-Hill's 500 College Algebra and Trigonometry Questions*. We are here to help you improve your grades on classroom, midterm, and final exams. These 500 questions will help you study more effectively, use your preparation time wisely, and get the final grade you want.

This book gives you 500 multiple-choice questions that cover the most essential course material. Each question has a detailed answer explanation. These questions give you valuable independent practice to supplement your regular textbook and the groundwork you are already doing in the classroom.

You might be the kind of student who needs to study extra questions a few weeks before a big exam for a final review. Or you might be the kind of student who puts off preparing until right before a midterm or final. No matter what your preparation style, you will surely benefit from reviewing these 500 questions that closely parallel the content, format, and degree of difficulty of the questions found in typical college-level exams. These questions and their answer explanations are the ideal last-minute study tool for those final days before the test.

Remember the old saying "Practice makes perfect." If you practice with all the questions and answers in this book, we are certain that you will build the skills and confidence that are needed to ace your exams. Good luck!

—Editors of McGraw-Hill Education

McGraw-Hill's

500
College Algebra
and
Trigonometry
Questions

Review of Basic Algebra

Number Concepts

1. For each statement, tell which of the following properties or definitions are used.

Commutative	Additive Identity	Multiplicative Identity
Associative	Inverse	Division
Distributive	Subtraction	Negative

(A) $(-1)+[-(-1)]=0$

(B) $7 \div 9 = 7\left(\dfrac{1}{9}\right)$

(C) $1\left(-\dfrac{2}{3}\right) = -\dfrac{2}{3}$

(D) $8-12 = 8+(-12)$

(E) $7(s+t) = 7s+7t$

(F) $(3p+9)+3 = 3p+(9+3)$

(G) $x+ym = x+my$

2. Perform the indicated operation.

(A) $(-2)(3)(-5)$

(B) $40 + (-7)$

(C) $-8 - (-6) + 2$

(D) $\dfrac{3-\frac{2}{3}}{5+\frac{5}{6}}$

3. Perform the indicated operations. Write the answer in the form $a + bi$. (See Chapter 11 for a complete treatment of the complex numbers.)

(A) $(1 + i) + (3 - 2i)$

(B) $(2 - i) - (3 - 4i)$

(C) $(2 + i)(3 + 2i)$

(D) $\dfrac{2+i}{1-i}$

(E) $(2 + \sqrt{-9})(1 + \sqrt{-4})$

4. Prove that $i^{4k} = 1$ for all natural numbers k.

5. Solve for x and y: $3 - 2i = 4xi + 2y$.

Integral and Rational Exponents

6. Evaluate the given expression.

(A) $\dfrac{7^6}{7^4}$

(B) $3^{41} \cdot 3^{-9}$

(C) $\dfrac{2 + 2^{-1}}{5} + (-8)^0 - 4^{\frac{3}{2}}$

(D) $125^{-\frac{4}{3}}$

For questions 7–9, simplify the given expression.

7. $(2ab^2)^3(a^2c)^2$

8. $\dfrac{x^{-8} \cdot x^{-7}}{x^{-6}} \div \dfrac{x^{-5} \cdot x^{-4}}{x^{-3}}$

9. $\dfrac{\sqrt{a} \cdot a^{-\frac{2}{3}}}{\sqrt[6]{a^3}} + \dfrac{a^{-\frac{5}{6}}}{\sqrt[3]{a^2} \cdot a^{-\frac{1}{2}}}$

For questions 10–12, simplify and write in simplest radical form. Assume that all letters and radicands represent positive real numbers.

10. $2a\sqrt[3]{8a^8b^{13}}$

11. $\sqrt[3]{\dfrac{3y^5}{4x^4}}$

12. $\dfrac{2}{x^2 - \sqrt{x^4 + 2x^2 + 1}}$

Algebraic Expressions

For questions 13–18, perform the indicated operations and simplify.

13. $(x^2 - 2x + 3) + (4x^2 - x + 6)$

14. $(x^2 - 2x + 3) - (-5x^3 - 7x + 1)$

15. $(x - 2y)(x + 3y)$

16. $(\sqrt{s} + \sqrt{t})^2$

17. $(s + t)^3$

18. $m - \{m - [m - (m - 1)]\}$

19. Evaluate the polynomial $2x^2 - 3x - 10$ when $x = -5$.

20. Let n represent a number. Express each of the following algebraically.
   (A) 27 more than the number
   (B) The number increased by 45
   (C) 43 less than the number
   (D) 43 less the number

Factoring and Fractional Expressions

For questions 21–29, factor the given polynomial completely (relative to the integers).

21. $x^2 + 4x + 3$

22. $2x^2 + 7x + 6$

23. $4y^2 - 25$

24. $m^3 - 6m - 3$

25. $x^2 + x + 1$

26. $27p^3 + 8q^3$

27. $2x^4 - 24x^3 + 40x^2$

28. $4x^2 - 9y^2 + 4x + 1$

29. $a^4 - b^4$

For questions 30–38, perform the indicated operations and reduce to lowest terms. Notice that factoring is important in many of these problems.

30. $\dfrac{x^2 - 4}{x^2 - 4x - 12}$

31. $\dfrac{4m-3}{18m^3} + \dfrac{3}{m} - \dfrac{2m-1}{6m^2}$

32. $\dfrac{x^2}{12} + \dfrac{x}{18} - \dfrac{1}{30}$

33. $\dfrac{x^2 - 2x - 15}{x^3 - 3x^2 - 10x}$

34. $\dfrac{2-x}{2x+x^2} \cdot \dfrac{x^2 + 4x + 4}{x^2 - 4}$

35. $\dfrac{2}{x^2 - 4} + \dfrac{6}{x-2}$

36. $\dfrac{1}{x-1} + \dfrac{2x}{(x-1)^2} - \dfrac{4-x}{(x-1)^3}$

37. $\dfrac{2 + \frac{1}{t}}{2 - \frac{1}{t}}$

38. $\dfrac{x+y}{x^{-1} + y^{-1}}$

39. Rationalize the denominator and reduce to lowest terms.

$\dfrac{1 - \sqrt{x+1}}{1 + \sqrt{x+1}}$

40. Rationalize the numerator.

$\dfrac{\sqrt{2+h} + \sqrt{2}}{h}$

CHAPTER 2

Equations and Inequalities

Linear Equations

For questions 41–42, solve the given equation, if possible. In each case, the replacement set for the variable is the set of real numbers.

41. $6x + 3 = 19x + 5$

42. $2(3x - 6) + 4(3x - 5) = 14x$

For questions 43–49, solve for x or y and check the given equation.

43. $x - 2(1 - 3x) = 6 + 3(4 - x)$

44. $ay + b = cy + d$

45. $\dfrac{3x - 2}{5} = 4 - \dfrac{1}{2}x$

46. $\dfrac{3x + 1}{3x - 1} = \dfrac{2x + 1}{2x - 3}$

47. $\dfrac{1}{x - 3} - \dfrac{1}{x + 1} = \dfrac{3x - 2}{(x - 3)(x + 1)}$

48. $\dfrac{x^2 - 2}{x - 1} = x + 1 - \dfrac{1}{x - 1}$

49. $\dfrac{1}{x - 1} + \dfrac{1}{x - 3} + \dfrac{2x - 5}{(x - 1)(x - 3)}$

50. Find k if the given number is a solution of the given equation.
(A) $12,\ 2x + 5 = 3x + k$
(B) $2,\ x^2 + kx + 2 = 0$

51. The sum of two consecutive even integers is 10. Find the integers.

52. The perimeter of a rectangle is 30 m, and its length is twice its width. Find the length and width of the rectangle.

53. Train A leaves station Q at the same time as train B leaves station R. Train A travels at a rate of 30 miles per hour (mi/h) directly toward R, and B travels at 45 mi/h directly toward Q. How many miles must A travel before the trains meet if the stations are 60 miles apart? (See Figure 2.1.)

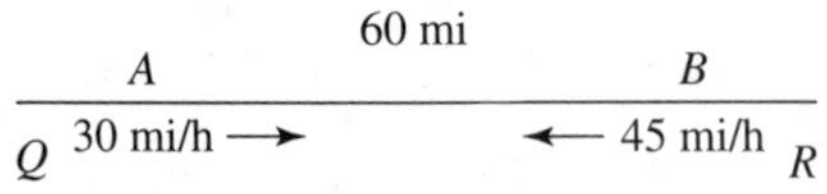

Figure 2.1

54. The sale price on a camera after a 20 percent discount is $72. What was the price before the discount?

55. Suppose Kate can paint a particular room in 6 h. At what minimum rate must her helper be able to paint the room alone if together they must complete the job in $3\frac{3}{7}$ h?

56. Barbara is twice as old as Mary, and Dick is three times as old as Barbara. Their average age is 36. How old is Barbara?

57. Juanita has nickels and dimes in her pocket. Their total value is $1, and there are twice as many dimes as nickels. How many nickels does she have?

Nonlinear Equations

For questions 58–67, solve the given equation. Leave any answer containing a radical in simplest radical form.

58. $x^2 = 9$

59. $t^2 + 5 = 0$

60. $4x^2 - 7 = 0$

61. $(n+5)^2 = 9$

62. $x^2 + 3x + 2 = 0$

63. $2t^2 - 2t = 12$

64. $x^2 + x - 1 = 0$

65. $x^2 + 4x + 1 = 0$

66. $(x^2 - 3)(x^2 - 4) = 0$

67. $8y^{-2} + 6y^{-1} + 1 = 0$

68. Find the discriminant.
(A) $Z^2 + 5Z - 6 = 0$
(B) $S^2 + 5S = 3$

69. Determine whether the equation has real roots.
(A) $x^2 + 11x + 11 = 0$
(B) $x^2 - 3x + \dfrac{9}{4} = 0$
(C) $2x^2 + x + 1 = 0$

70. Reduce the given equation to a quadratic and solve.

$$\sqrt{3w - 2} - \sqrt{w} = 2$$

71. The height (in feet) of a ball thrown vertically upward above the ground in t seconds (s) is given by $h = 128t - 16t^2$. In how many seconds will the ball be 192 ft high?

72. If P dollars are invested at r percent compounded annually, at the end of 2 years the amount will be $A = P(1 + r)^2$. At what interest rate will $1000 increase to $1400 in 2 years?

73. Find two consecutive positive integers whose product is 210.

Linear Inequalities

74. Write an algebraic expression for each statement.
   (A) $3y$ less than 4 times z is nonpositive.
   (B) 5 times t is less than or equal to 3 less than 3 times negative y.
   (C) 5 times t is more than 3 more than 3 times y.
   (D) 5 times t is 3 more than 3 times y.
   (E) 5 times t is more than 3 times y.

75. Is the given number a solution of the given inequality?
   (A) $-x + 5 \geq 0; 3$
   (B) $\dfrac{4}{x} + 3 \geq \dfrac{1}{x}; \dfrac{1}{2}$
   (C) $x^{-1} + 1 < x^{-2} - 2; 1$

For questions 76–78, solve the given inequality (if possible).

76. $14t \leq 3t - 4$

77. $-3s > 5s + 2$

78. $-6 < x + 2 < 10$

For questions 79–80, an inequality is given. Graph the solution, and indicate the solution using interval notation.

79. $14t \leq -3t - 4$

80. $-6 < x + 2 < 10$

81. $a, b < 0$ and $a > b$. Tell whether each statement is true or false. If it is false, give a counterexample.
   (A) $a^2 > b^2$
   (B) $ab < b^2$
   (C) $a(a - b) > b(a - b)$
   (D) $ab(a - b) > 0$
   (E) $\dfrac{(a - b)^3}{b} > 0$

82. Let a and b be positive integers whose sum is less than 12.
   (A) If we know that a is either 1 or 2, how many possible choices are there for b?
   (B) How many possibilities are there altogether for a and b?

Absolute Value

For questions 83–89, solve each equation or inequality, if possible.

83. $|x| = 6$

84. $|x| < 5$

85. $|x| < -10$

86. $|x - 4| = 3$

87. $|t - 2| \geq 5$

88. $|2x - 5| < 10$

89. $|1 - x| < 5$

Nonlinear Inequalities

For questions 90–92, solve the given inequality. Graph the solution set for each even-numbered problem, and express each odd-numbered solution set in interval form.

90. $x^2 - 3x + 2 > 0$

91. $3x^2 + x > 0$

92. $3x^2 < -2x + 1$

Miscellaneous Problems

93. Solve for $x : \sqrt{5x - 1} - \sqrt{x} = 1$

94. Determine the character of the roots.
   (A) $x^2 - 8x + 9 = 0$
   (B) $3x^2 - 8x + 9 = 0$
   (C) $6x^2 - 5x - 6 = 0$
   (D) $4x^2 - 4\sqrt{3}x + 3 = 0$

95. Two pipes together can fill a reservoir in 6 h 40 min. Find the time each alone will take to fill the reservoir if one of the pipes can fill it in 3 h less time than the other.

96. Express $x^2 + y^2 - 6x - 9y + 2 = 0$ in the form $a(x-h)^2 \pm b(y-k)^2 = c$.

97. Transform each of the following into the form $a\sqrt{(x-h)^2 + k}$ or $a\sqrt{k-(x-h)^2}$.

    (A) $\sqrt{4x^2 - 8x + 9)}$

    (B) $\sqrt{8x - x^2}$

    (C) $\sqrt{3 - 4x - 2x^2}$

98. If an object is thrown directly upward with initial speed v feet per second (ft/s), its distance s ft above the ground after t s is given by $s = vt - \frac{1}{2}gt^2$. Taking $g = 32.2$ ft/s^2 and the initial speed as 120 ft/s, find out

    (A) when the object is 60 ft above the ground.

    (B) when it is highest in its path and how high it is.

99. Without sketching, state whether the graph of each of the following functions crosses the x axis, is tangent to it, or lies wholly above or below it.

    (A) $3x^2 + 5x - 2$

    (B) $2x^2 + 5x + 4$

    (C) $4x^2 - 20x + 25$

    (D) $2x - 9 - 4x^2$

100. Form the quadratic equation whose roots x_1 and x_2 are

    (A) $3, \dfrac{2}{5}$

    (B) $-2 + 3\sqrt{5}, -2 - 3\sqrt{5}$

    (C) $\dfrac{3 - i\sqrt{2}}{2}, \dfrac{3 + i\sqrt{2}}{2}$

101. Determine k so that the given equation will have the stated property, and write the resulting equation.

    (A) $x^2 + 4kx + k + 2 = 0$ has one root 0.

    (B) $4x^2 - 8kx - 9 = 0$ has one root the negative of the other.

    (C) $4x^2 - 8kx + 9 = 0$ has roots whose difference is 4.

CHAPTER 3

Graphs, Relations, and Functions

Cartesian Coordinate System

102. Plot each of the following points in the Cartesian coordinate system: $A(2, 0)$, $B(-2, 5)$, $C(-3, -2)$, $D(2, 1)$.

For questions 103 and 104, find the distance between the given points.

103. $A(1, 0)$ and $B(0, 1)$

104. $A(1, 4)$ and $B(-1, 5)$

105. Find the midpoint of the segment joining the two given points $(1, 2)$ and $(3, 2)$.

106. Decide whether the graphs of the following equations exhibit symmetry with respect to the x axis or origin. Do not graph the equations to determine whether the symmetry exists.

The tests for symmetry are the following:

- Replace x with $-x$. If the equation *does not* change, the graph is symmetric with respect to the y axis.
- Replace y with $-y$. If the equation *does not* change, the graph is symmetric with respect to the x axis.
- Replace x with $-x$ and y with $-y$. If the equation *does not* change, the graph is symmetric with respect to the origin.
  - (A) $y = x$
  - (B) $x^2 + y^2 = 4$
  - (C) $y = |2x|$
  - (D) $|x| + 1 = y$
  - (E) $y = x^3$

107. Graph $y = |x| + 1$.

Relations and Functions

108. Tell whether the relation shown is a function. In all cases, x is the independent variable.

(A) See Figure 3.1.

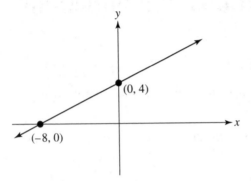

Figure 3.1

(B) See Figure 3.2.

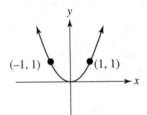

Figure 3.2

(C) See Figure 3.3.

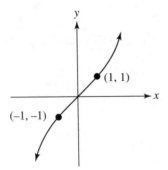

Figure 3.3

(D) See Figure 3.4.

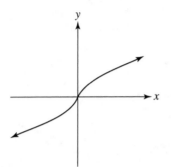

Figure 3.4

(E) See Figure 3.5.

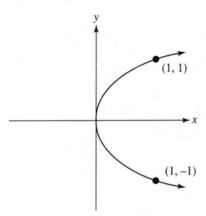

Figure 3.5

109. Tell whether the equation defines a function.

(A) $f(x) = 2x - 10$
(B) $f(x) = -2x^2 + 8$
(C) $y = |x|$
(D) $x = |y|$
(E) $x = 2y^2$

110. Find the indicated value if $f(x) = x - 9$, $g(x) = x^2 - 9$, and $F(x) = x^3 - x + 4$.

(A) $F(2) - g(3)$
(B) $F(1) + F(2)$
(C) $F(0) \cdot f(0)$
(D) $\dfrac{f(0)}{F(0)}$

111. Find the domain of the given relation. In all cases, x is the independent variable.

(A) $2x + 5 = y$

(B) $y = x^2$

(C) $y = \sqrt{3 - x^2}$

(D) $y = |x| - 1$

(E) $y = \dfrac{3}{2 - x}$

(F) $x^2 + y^2 = 8$

(G) $\{(x, y) \mid x \in \mathcal{R}, y \in \mathcal{R}, y = 3\}$

(H) $\{(1, 4), (2, 2), (3, 8)\}$

112. Find the range of the given function.

(A) $\{(1, 0), (0, 1), (2, a)\}$

(B) $f(x) = 2x - 3$

(C) $f(x) = x^2$

(D) $g(x) = 2 - |x|$

(E) $f(x) = |x| + 5$

(F) $f(x) = \sqrt{2 - x}$

(G) $h(x) = 4$

(H) $f(x) = \dfrac{1}{x}$

113. Tell whether the function is one-to-one.

(A) $f(x) = 3x + 4$

(B) $f(x) = |x| + 1$

(C) $g(x) = 3\sqrt{1 - x}$

(D) $f(x) = \{(1, 2), (2, 1)\}$

(E) $h(x) = \{(1, 3), (2, 4), (3, 1), (3, 4)\}$

Graph of a Function

114. For questions 114A–114D, refer to Figure 3.6, which is the graph of $y = f(x)$.

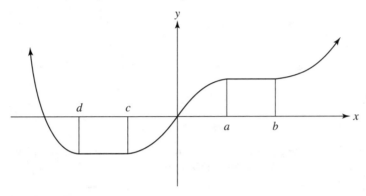

Figure 3.6

(A) Is the relation $y = f(x)$ a function?
(B) Find all intervals over which $y = f(x)$ is constant.
(C) Find all intervals over which $y = f(x)$ is increasing and all intervals over which it is decreasing.
(D) Find all intervals over which $y = f(x)$ is nonincreasing and those over which it is nondecreasing.

For questions 115–120, graph the given function.

115. $f(x) = 2x - 3$

116. $f(x) = x^2$

117. $f(x) = x^3$

118. $f(x) = \sqrt{x+2}$

119. $f(x) = |x|$

120. $f(x) = [x]$

Step Functions and Continuity

For questions 121 and 122, graph the given function.

121. $h(x) = \begin{cases} 1 & x > 1 \\ 2 & -2 < x \le 1 \\ -1 & x \le -2 \end{cases}$

122. $f(x) = \begin{cases} x & x > 2 \\ x^2 & x \le 2 \end{cases}$

123. Tell whether the given function is continuous for every x in its domain. If it is not, tell where it is discontinuous.

(A) $f(x) = x^2$

(B) $f(x) = x^3 - 1$

(C) $g(x) = ax^2 + bx + c; \, a, b, c \in \mathcal{R}; \, b \ne 0.$

(D) $f(x) = \begin{cases} 1 & x \ge 1 \\ 0 & x < 1 \end{cases}$

Linear Functions

For questions 124–127, find the slope and y intercept (if they exist).

124. $y = 3x + 1$

125. $-3y = x + 6$

126. $y = 3$

127. $x = -2$

For questions 128 and 129, graph the linear functions given. Use the slope and y intercept graphing technique.

128. $y = 3x + 2$

129. $y = -x + 1$

For questions 130–135, find an equation of the line that satisfies the given conditions.

130. No slope, x intercept is 7

131. Contains the points $(1, -1)$ and $(2, 3)$

132. Contains the point $(2, 1)$ and is parallel to $y = x - 3$

133. Contains the point $(3, 1)$ and is perpendicular to $2x + y = 3$

134. Vertical line containing $(7, -6)$

135. Horizontal line containing $(7, -6)$

136. Find the point of intersection of $2x + 1 = y$ and $x - y = 3$.

137. Find all points of intersection of $y = x^2$ and $y = x$.

Algebra of Functions

For questions 138–140, find $(f + g)(x)$, $(f - g)(x)$, and their respective domains.

138. $f(x) = x^2$, $g(x) = x - 1$

139. $f(x) = \dfrac{1}{x-1}$, $g(x) = \sqrt{x}$

140. $f(x) = \sqrt{x+1}$, $g(x) = \sqrt{x-1}$

141. $f(x) = \sqrt{x+1}$, $g(x) = x - 3$. Find $fg(x)$ and $\dfrac{f}{g}(x)$ and their respective domains.

For questions 142 and 143, find $f \circ g(x)$.

142. $f(x) = x^2$, $g(x) = 2x + 5$

143. $f(x) = \dfrac{1}{1+x}$, $g(x) = \dfrac{1}{2+x}$

For questions 144 and 145, find $g \circ f(x)$. These are functions from questions 142 and 143 above. Check the results to notice that $f \circ g(x)$ and $g \circ f(x)$ are not always the same function.

144. $f(x) = x^2$, $g(x) = 2x + 5$

145. $f(x) = \dfrac{1}{1+x}$, $g(x) = \dfrac{1}{2+x}$

146. Let $f(x) = 2x - 4$.

    (A) Find the domain and range of f and f^{-1}.

    (B) Find $f^{-1}(x)$.

    (C) Find $f^{-1} \circ f(x)$.

    (D) Find $f \circ f^{-1}(x)$.

    (E) Graph $f(x)$ and $f^{-1}(x)$ in the same coordinate system. Show that these two graphs are symmetric about $y = x$.

147. Let $f(x) = x^2 + 1$, $x \geq 0$.

    (A) Find the domain and range of f^{-1}.

    (B) Find $f^{-1}(x)$.

    (C) Find $f^{-1} \circ f(x)$.

    (D) Find $f \circ f^{-1}(x)$.

    (E) Graph $f(x)$, $f^{-1}(x)$, and $y = x$ on the same axis system.

Problem Solving and Formulas

148. An open box is to be formed from a rectangular sheet of tin 20×32 in by cutting equal squares, x in on a side, from the four corners and turning up the sides. Express the volume of the box as a function of x.

149. A farmer has 600 ft of woven wire fencing available to enclose a rectangular field and to divide it into three parts by two fences parallel to one end. If x ft of stone wall is used as one side of the field, express the area enclosed as a function of x when the dividing fences are parallel to the stone wall. Refer to Figure 3.7.

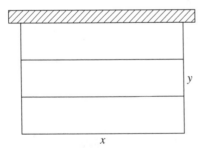

Figure 3.7

150. A right circular cylinder is said to be inscribed in a sphere if the circumference of the base of the cylinder is in the surface of the sphere. If the sphere has radius R, express the volume of the inscribed right circular cylinder as a function of the radius r of its base.

151. Let $z = f(x, y) = 2x^2 + 3y^2 - 4$. Find:

(A) $f(0, 0)$
(B) $f(2, -3)$
(C) $f(-x, -y)$

CHAPTER 4

Polynomial and Rational Functions

Polynomial Functions

152. Write the given equation in standard form.

    (A) $4x^2 + 2x^3 - 6 + 5x = 0$

    (B) $2x^5 + x^3 + 4 = 0$

    (C) $\dfrac{x^5}{2} - 1 = 0$

    (D) $(x + 2)^2 + 5 = 0$

For questions 153 and 154, find the vertex of the given parabola.

153. $y + 1 = 2(x - 3)^2$

154. $y = x^2 - 4x + 6$

155. Identify the given parabola as opening upward or downward.

    (A) $y = x^2 - 5x + 6$

    (B) $y = 2 - x^2$

    (C) $y = x(1 - x)$

    (D) $y = (2 - x)(3 - x)$

For questions 156 and 157, find the maximum (or minimum) value of the quadratic function.

156. $f(x) = 2(x - 3)^2 - 1$

157. $f(x) = -2x^2 + 3$

158. Is the given function a polynomial function?

(A) $y = x^2$

(B) $y = \dfrac{1}{x^2}$

(C) $y = (x + 2)^x$

(D) $y = 2x + 4 - 2i$

(E) $y = 2x^2 + \sqrt{3}x - i$

(F) $y = ix^3 - \sqrt{4x} - 2$

For questions 159 and 160, determine whether the graph of the polynomial will cross or touch the x axis at each zero.

159. $f(x) = x^2(x - 1)$

160. $f(x) = (x^2 - 1)(x + 2)^3$

161. Find the remainder by using the remainder theorem.

$(x^3 - 5x^2 - 3x + 15) \div (x + 2)$

162. Show that the second expression is a factor of the first by means of the factor theorem.

$2x^3 - 3x^2 + 2x - 8, \ x - 2$

163. Write an equation with integral coefficients having the given numbers and no others as roots.

1, 2, –3

Graphing Polynomial Functions

For questions 164–166, sketch the graph of the given polynomial function.

164. $f(x) = x^3$

165. $y = (x - 3)(x^2 - 1)$

166. $y = (x^2 - 1)(x^2 - 9)$

167. Refer to the following situation: A parcel delivery service will deliver only packages with length plus girth not exceeding 108 in. A packaging company wishes to design a box with a square base that will have a maximum volume and will meet the delivery service's restrictions.

(A) Write the volume of the box $V(x)$ in terms of x.
(B) What is the domain of V in question 167A?
(C) Graph V for the domain in question 167B.
(D) From the graph in the previous question (see Figure A4.4 in the Answers), estimate (to the nearest inch) the dimensions of the box with maximum volume. What is this maximum volume?

168. Refer to the graphs in Figures 4.1–4.6.

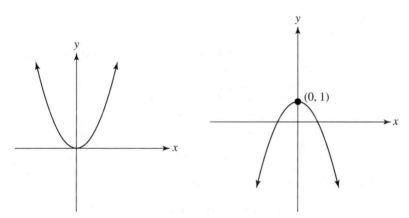

Figure 4.1   **Figure 4.2**

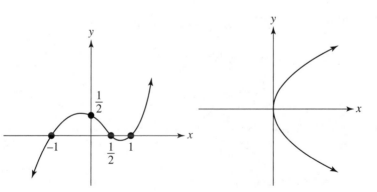

Figure 4.3   **Figure 4.4**

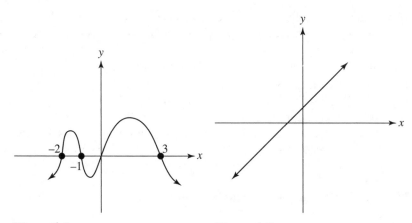

Figure 4.5   **Figure 4.6**

(A) Which of the figures represents a polynomial function which is always increasing?
(B) Which of the figures are symmetric about the y axis?
(C) Which of the figures exhibits x axis symmetry?
(D) Which figure has four zeros?
(E) In which figures are there two intervals in which the function is positive?

Synthetic Division

169. Use synthetic division to find the quotient and remainder.

$$(x^3 + x^2 + x + 3) \div (x - 1)$$

For questions 170 and 171, evaluate by synthetic division.

170. $f(-2)$ if $f(x) = x^3 - 7x^2 + 12x - 3$

171. $f(3)$ if $f(x) = x^3 - 7x^2 + 12x - 3$

For questions 172 and 173, use synthetic division to show that the first polynomial is a factor of the second.

172. $x - 3, x^3 - 18x + 27$

173. $x + \dfrac{1}{2}, x^4 - \dfrac{3x^2}{4} + \dfrac{x}{2} + \dfrac{3}{8}$

Fundamental Theorem of Algebra

174. State the fundamental theorem of algebra.

For questions 175 and 176, find a polynomial equation of lowest degree which has the roots listed.

175. 0, 1, 2

176. $1+\sqrt{3}, 1-\sqrt{3}, -1-\sqrt{3}$

177. $(2-3i), (2+3i), -4$ (multiplicity = 2) Find a polynomial $P(x)$ of lowest degree with leading coefficient 1 (monic) that has the indicated set of zeros. You may leave the answer in factored form, but indicate the degree of the polynomial.

For questions 178–180, find the zeros of the given polynomial.

178. $x^2(x-2)(x+3) = 0$

179. $(x+i)^2(x-1)^3(x+4) = 0$

180. $(2x-1)(2x+5)(x-3) = 0$

For questions 181–183, find the remaining zeros of each given polynomial, using the given zero(s).

181. $x^3 - 3x^2 + x + 1$, 1

182. $x^3 - 3x^2 + 2$, $1 + \sqrt{3}$

183. $x^4 - x^3 + 6x^2 - 26x + 20$, $-1 - 3i$

184. Find all possible candidates for rational zeros of the given polynomial.
$2x^2 - 5x + 2$

For questions 185 and 186, find the rational zeros of the given polynomial.

185. $x^3 + 3x^2 + x - 2$

186. $2x^3 - x^2 - 5x - 2$

187. Find all roots for $x^4 - x^3 - 5x^2 + 3x + 2$.

For questions 188 and 189, use Descartes' rule of signs to discuss the number of positive and negative zeros of each given polynomial.

188. $P(x) = 2x^2 + x - 4$

189. $S(x) = x^5 + x^4 + x^3 - x^2 + 1$

190. For the given equation, show that there is at least one real zero between the given values of a and b.

$P(x) = x^2 - 3x - 2; a = 3, b = 4$

Rational and Algebraic Functions

191. Tell whether the given function is rational and why?

(A) $f(x) = cx + d$

(B) $f(x) = \dfrac{(x-2)^{\frac{3}{2}}}{x-2}$

(C) $g(x) = |x|$

(D) $f(x) = \dfrac{1}{\pi x}$

(E) $h(x) = 3^{x^2} + 2$

(F) $y = \dfrac{x^2 + 3x - 7}{x^2 + 5x - 6}$

(G) $f(x) = (x + 2)^x$

For questions 192–194, find all vertical and horizontal asymptotes of the graph of the given function. Do *not* graph the function.

192. $f(x) = \dfrac{x}{x+1}$

193. $f(x) = \dfrac{x^2}{x^2 - 4}$

194. $f(x) = \dfrac{x+3}{(x-2)(x+1)(x+2)}$

For questions 195 and 196, sketch the graph of the given rational function.

195. $f(x) = \dfrac{x}{x+1}$

196. $f(x) = \dfrac{x^2}{x^2 - 4}$

CHAPTER 5

Systems of Equations and Inequalities

Systems of Linear Equations

197. Solve the system of equations by graphing.

$x + 2y = 5$, $3x - y = 1$

198. Without solving, tell whether the given system is independent, dependent, or inconsistent.

(A) $3x + y = 6$, $2x + y = 5$
(B) $3x - y = 4$, $x + y = 7$
(C) $2x + y = 10$, $-4x - 2y = -20$
(D) $2x - y = -10$, $-4x - y = 10$
(E) $y + x = 5$, $y + x = 8$
(F) $x = 3y + 8$, $2y = x + 4$
(G) $y = x$, $2y = x + 6$

For questions 199–201, solve by using the addition/subtraction method.

199. $4x - y = 10$, $x + 3y = 9$

200. $2x + y = 5$, $4x + 2y = 10$

201. $3x + y = 20$, $3x + y = 30$

202. Solve by substitution.

$x + y = 16$, $x - y = 10$

For questions 203–206, solve the given problem by using a system of linear equations.

203. The sum of two numbers is 20, and their difference is 10. Find the two numbers.

204. The length of a rectangle is 10 in more than the width. The perimeter of the rectangle is 60 in. Find the dimensions of the rectangle.

205. If the numerator of a fraction is increased by 2, the fraction is $\frac{1}{4}$; if the denominator is decreased by 6, the fraction is $\frac{1}{6}$. Find the fraction.

206. A woman can row downstream 6 mi in 1 h and return in 2 h. Find her rate in still water and the rate of the river.

207. Solve the system $\begin{cases} x - 5y + 3z = 9 & (1) \\ 2x - y + 4z = 6 & (2) \\ 3x - 2y + z = 2 & (3) \end{cases}$

Matrices and Determinants

208. Find a, b, c, d. $\begin{bmatrix} a & b \\ c & d \end{bmatrix} = \begin{bmatrix} 1 & 5 \\ -2 & 3 \end{bmatrix}$

For questions 209–213, refer to the following matrices.

$$A = \begin{bmatrix} 1 & -1 \\ 3 & 1 \end{bmatrix} \quad B = \begin{bmatrix} -3 & 2 \\ -2 & -3 \end{bmatrix} \quad C = \begin{bmatrix} -2 \\ -3 \\ 1 \end{bmatrix}$$

$$D = \begin{bmatrix} 2 \\ 3 \\ 5 \end{bmatrix} \quad E = \begin{bmatrix} -4 & -1 & 0 & -2 \end{bmatrix} \quad F = \begin{bmatrix} -2 & -3 \\ -2 & 0 \\ 1 & -2 \\ 3 & -5 \end{bmatrix}$$

209. What are the dimensions of B and of E?

210. What element is in the second row, first column of F?

211. Find $A + B$.

212. Find the negative of matrix B.

213. Find $3D - 4C$.

For questions 214 and 215, let $A = \begin{bmatrix} 2 & 3 & 1 \\ 0 & -4 & 5 \end{bmatrix}$ and $B = \begin{bmatrix} -5 & 2 & 4 \\ 3 & 0 & -1 \end{bmatrix}$.

214. Find $2B$.

215. Find A^T.

For questions 216–220, find the product indicated.

216. $\begin{bmatrix} 1 & 2 \\ 3 & 1 \end{bmatrix} \begin{bmatrix} -1 & 2 \\ -1 & 2 \end{bmatrix}$

217. $\begin{bmatrix} 1 \\ 1 \\ 0 \end{bmatrix} \begin{bmatrix} 2 & 3 & 1 \end{bmatrix}$

218. $\begin{bmatrix} 2 & 1 & 3 \end{bmatrix} \begin{bmatrix} 2 & 1 & 0 \\ 0 & 0 & 0 \\ 1 & 0 & 2 \end{bmatrix}$

219. $\begin{bmatrix} 4 \\ 1 \\ 6 \end{bmatrix} \begin{bmatrix} 1 & 0 & 1 \\ 0 & 2 & 0 \\ 0 & 0 & 1 \end{bmatrix}$

220. $\begin{bmatrix} 1 & 2 & 3 \\ 1 & 0 & 1 \end{bmatrix} \begin{bmatrix} 3 & 1 \\ 4 & 2 \\ 6 & 0 \end{bmatrix}$

221. Perform the indicated operations. The · represents the *dot product*.

$$\begin{bmatrix} 1 & 6 \end{bmatrix} \cdot \begin{bmatrix} 0 \\ 1 \end{bmatrix} + \begin{bmatrix} 2 & 3 & 4 \end{bmatrix} \cdot \begin{bmatrix} 1 \\ 2 \\ 3 \end{bmatrix}$$

For questions 222–225, verify the given statements,

where $A = \begin{bmatrix} 1 & 2 \\ 0 & 1 \end{bmatrix}$, $B \begin{bmatrix} 1 & 1 \\ 2 & 3 \end{bmatrix}$, $C \begin{bmatrix} -3 & 1 \\ -1 & 2 \end{bmatrix}$.

222. $AB \neq BA$.

223. $(AB)C = A(BC)$.

224. $A(B + C) = AB + AC$.

225. $(B + C)A = BA + CA$.

226. Evaluate the determinant.

$$\begin{vmatrix} 1 & 6 \\ 5 & 4 \end{vmatrix}$$

227. Evaluate the determinant by using cofactors.

$$\begin{vmatrix} 1 & 4 & 1 \\ 1 & 1 & -2 \\ 2 & 1 & -1 \end{vmatrix}$$

228. State the theorem that can be used to justify the given statement.

(A) $\begin{vmatrix} 6 & 8 \\ 0 & -1 \end{vmatrix} = 2 \begin{vmatrix} 3 & 4 \\ 0 & -1 \end{vmatrix}$

(B) $\begin{vmatrix} 1 & 3 & 6 \\ 0 & 0 & 0 \\ 4 & 9 & 1 \end{vmatrix} = 0$

(C) $\begin{vmatrix} 0 & 4 & 6 & 3 & 8 \\ 0 & 1 & 2 & 1 & 0 \\ 0 & 1 & 6 & 18 & 9 \\ 0 & 1 & 4 & 0 & 1 \\ 0 & 0 & 1 & 1 & 0 \end{vmatrix} = 0$

(D) $\begin{vmatrix} 4 & 3 \\ 1 & 2 \end{vmatrix} = \begin{vmatrix} 4-3 & 3-6 \\ 1 & 2 \end{vmatrix}$

(E) $\begin{vmatrix} 1 & 2 & 4 & 1 \\ 1 & 3 & 4 & 2 \\ 1 & 2 & 4 & 1 \\ 0 & 1 & 4 & 1 \end{vmatrix} = 0$

(F) $\begin{vmatrix} 2 & 1 & 3 \\ 4 & 6 & 9 \\ 2 & 8 & 1 \end{vmatrix} = - \begin{vmatrix} 4 & 6 & 9 \\ 2 & 1 & 3 \\ 2 & 8 & 1 \end{vmatrix}$

229. Show that the two matrices are inverses of each other.

$$\begin{bmatrix} 3 & -4 \\ -2 & 3 \end{bmatrix} \begin{bmatrix} 3 & 4 \\ 2 & 3 \end{bmatrix}$$

For questions 230 and 231, a matrix M is given. Find M^{-1}.

230. $\begin{bmatrix} 1 & 2 \\ 1 & 3 \end{bmatrix}$

231. $\begin{bmatrix} 1 & -3 & 0 \\ 0 & 3 & 1 \\ 2 & -1 & 2 \end{bmatrix}$

For questions 232 and 233, use Cramer's rule to solve the given system.

232. $x + 2y = 6,\ 3x - 5y = 10$

233. $x + y = 0,\ 2y + z = -5,\ -x + z = -3$

Systems of Nonlinear Equations

234. Solve the system algebraically.

$$3x - y = 8,\ 3x^2 - y^2 = 26$$

235. Solve the system $\begin{cases} x^2 + y^2 = 25 & (1) \\ \quad\ xy = 12 & (2) \end{cases}$

For questions 236–240, solve the given problem by using a system of equations.

236. Two numbers differ by 2, and their squares differ by 48. Find the numbers.

237. The sum of the circumferences of two circles is 88 in, and the sum of their areas is $\frac{2200}{7}$ in^2 when $\pi \approx \frac{22}{7}$. Find the radius of each circle.

238. A party costing $30 is planned. It is found that by adding 3 more to the group, the cost per person would be reduced by 50¢. For how many people was the party originally planned?

239. The square of a certain number exceeds twice the square of another number by 16. Find the numbers if the sum of their squares is 208.

240. The diagonal of a rectangle is 85 ft. If the short side is increased by 11 ft and the long side decreased by 7 ft, the length of the diagonal remains the same. Find the original dimensions.

Systems of Inequalities

241. Graph the given inequality.

$$2x - 3y < 6$$

For questions 242 and 243, find the solution set of each system graphically.

242. $-2 \leq x < 2, -1 < y \leq 6$

243. $y \geq x^2, x \geq 0$

244. Find a parametric representation for the line segment $\overline{P_1 P_2}$.

$P_1(2, 3), P_2(5, 8)$

245. Find the maximum and minimum values for $f(t)$ in the given range.

$f(t) = 2t + 5, 0 \leq t \leq 4$

246. Express $f(x, y)$ as $g(t)$ for $\overline{P_1 P_2}$, and then find the extrema.

$f(x, y) = 3x + 2y - 5; P_1 = (2, 1), P_2(8, 6)$

CHAPTER 6

Exponential and Logarithmic Functions

Exponential Functions

247. Sketch $y = 2^x$ and $y = 2^{-x}$ on the same axes.

248. Sketch $y = e^x$.

249. Let $f(x) = a^x$.

    (A) Prove that $f(x + y) = f(x)f(y)$.

    (B) Prove that $f(x - y) = \dfrac{f(x)}{f(y)}$.

    (C) Prove that f is one-to-one.

    (D) Prove that $f(-x) = \dfrac{1}{f(x)}$.

    (E) Prove that $f(b + x) = a^b f(x)$.

250. Compute the compound amount. $2000 at 12 percent compounded semiannually for 3 years.

251. Compute the principal P invested to yield the following compound amounts A. $5000 at 10 percent compounded annually for 5 years.

252. Compute the amount due, given that the interest is compounded continuously. $3000 at 10 percent for 5 years.

Logarithmic Functions

For questions 253–257, solve the equation.

253. $\log 100 = x$

254. $\log x = 2$

255. $\log_x 81 = 4$

256. $\ln e^{x+2} = 7$

257. $\log_{e^2} x = 10$

258. Evaluate the given expression where $f(x) = \log x$, $g(x) = 10^x$, $h(x) = \ln x$, $k(x) = e^x$, $l(x) = x^2$.
  (A) $f \circ g(x)$
  (B) $g \circ f(x)$
  (C) $h \circ k(x)$
  (D) $f \circ l(10)$
  (E) $l \circ h(3)$

259. Sketch $y = \log_2 x$, $y = 2^x$ on the same axes.

260. Prove that $g(x) = \log_a x$ is one-to-one.

261. Find the domain and range of the given function.
  (A) $y = \log_3 (x + 1)$
  (B) $y = \log_3 (2x - 5)$
  (C) $y = \log_5 (x^2 + 1)$
  (D) $y = |\log_6 x|$

Properties of Log Functions

262. Write each expression as the algebraic sum of logarithms. The base is any positive real number except 1.
  (A) $\log (251)(46)(18)$
  (B) $\log (34)^2(2.7)$
  (C) $\log (24)^{\frac{1}{2}} (35)^3$
  (D) $\log \dfrac{(83)(41)}{29}$
  (E) $\log a^n b^m$
  (F) $\log \sqrt[n]{a^{n-1} p}$

263. Obtain the required logarithm for the following.

(A) $\log_2 (8)(16{,}384)$

(B) $\log_2 (16{,}384)^{-2}$

(C) $\log_2 \sqrt[4]{65{,}536}$

264. Write each expression in terms of a single logarithm with a coefficient of 1.

(A) $2 \log_b x - \log_b y$

(B) $3 \log_b x + 2 \log_b y - 4 \log_b z$

(C) $\dfrac{1}{5}(2 \log_b x + 3 \log_b y)$

(D) $\dfrac{1}{3}(\log_b x - \log_b y)$

265. Rewrite the given expression in terms of common logarithms.

(A) $\log_6 7$

(B) $\log_{\frac{1}{3}} 30$

(C) $\log_{20} e$

(D) $\log_{20} e^3$

266. Tell whether the given statement is true or false, and explain your answer.

(A) $(\log 15)^\circ = 1$

(B) If $f(x) = \log_8 x$, then the range of f is $\mathcal{R}$.

(C) The domain of $y = \log_x b$ is $\mathcal{R}$.

(D) $\log_x a = \log_{\frac{1}{x}}\left(\dfrac{1}{a}\right)$

(E) $f(x) = \log_b x$ is an always-decreasing function.

(F) $\log_k ab = (\log_k a)(\log_k b)$

(G) Let $g(x) = \log_a x$ and $h(x) = \log_b x$; then $g \circ h(ab) = \log_a (1 + \log a)$.

(H) Using functions g and h in question 266(G), $g \circ h(ab) = \log_a 1 + \log_a (\log_b a)$.

(I) $\log_a \left(\dfrac{1}{x}\right) = -\log_a x$

(J) $\log_b a = \dfrac{1}{\log_a b}$

267. Find x so that $\frac{3}{2}\log_b 4 - \frac{2}{3}\log_b 8 + 2\log_b 2 = \log_b x$.

268. Let $f(x) = 7^x$. Find the domain and range of $f^{-1}(x)$.

269. Solve the equation $A = Pe^{rt}$ for r.

Logarithmic and Exponential Equations

270. Solve the given equation for x in terms of y.

(A) $y = 10^x$

(B) $y = 3(10^{2x})$

(C) $y = \dfrac{e^x + e^{-x}}{2}$

For questions 271 – 276, solve the given equation; when appropriate, give answers to three decimal places.

271. $5^{x^2 - 3x} = 625$

272. $195^x = 2.68$

273. $\log_5 (x - 1) + \log_5 (x + 3) = 1$

274. $\log (3x + 4) = \log (5x - 6)$

275. $2 \ln 5x = 3 \ln x$

276. $\log x = \ln e$

CHAPTER 7

Trigonometric Functions

Angle Measurement

For questions 277–279, sketch the angle of each of the following measures. In your sketch, place the angle in standard position.

277. 225°

278. 420°

279. $-\dfrac{\pi}{3}$

280. $\dfrac{114°\,29'\,46''}{-81°\;\;4'\,11''}$

281. Express the given angle in degrees, minutes, and seconds.

40.25°

282. Rewrite the given angle in decimal form.

27°15′25″

For questions 283–286, convert the given angle to degrees if it is in radians and to radians if it is in degrees.

283. $\dfrac{\pi}{4}$

284. $-\dfrac{\pi}{6}$

285. 45°

286. $-135°$

287. Find the radian measure of a central angle θ subtended by an arc s in a circle of radius R, where R and s are given as shown.

$R = 4$ cm, $s = 24$ cm

288. Find five angles coterminal with $-155°$.

289. The minute hand of a clock is 12 in long. How far does the tip of the hand move during 20 min?

290. A railroad curve is to be laid out on a circle. What radius should be used if the track is to change direction by $25°$ in a distance of 120 ft?

291. Assuming the earth to be a sphere of radius 3960 mi, find the distance of a point in latitude $36°N$ from the equator.

For questions 292 and 293, in which quadrant will θ terminate?

292. $\sin \theta$ and $\cos \theta$ are both negative.

293. $\sec \theta$ is negative, and $\tan \theta$ is negative.

For questions 294 and 295, in which quadrants may θ terminate?

294. $\sin \theta$ is positive.

295. $\tan \theta$ is negative.

296. Determine the values of $\cos \theta$ and $\tan \theta$ if $\sin \theta = \dfrac{m}{n}$, a negative fraction.

297. Evaluate $\sin 0° + 2 \cos 0° + 3 \sin 90° + 4 \cos 90° + 5 \sec 0° + 6 \csc 90°$.

Trigonometric Functions

For questions 298–305, find the value of the trigonometric function at the given angle. Do *not* use a calculator.

298. $\sin 90°$

299. $\sin 45°$

300. $\cos 60°$

301. $\sin 120°$

302. $\cos 135°$

303. $\sin\left(\dfrac{3\pi}{4}\right)$

304. $\sec\left(\dfrac{4\pi}{3}\right)$

305. $\sin(-960°)$

For questions 306 and 307, evaluate the given expression.

306. $\sin^2 315° + \cos^2 315°$

307. $\cos\left(\dfrac{\pi}{4}\right)\cos\left(\dfrac{\pi}{2}\right) - \sin\left(\dfrac{\pi}{4}\right)\sin\left(\dfrac{\pi}{2}\right)$

For questions 308 and 309, find the value of the other five trigonometric functions for the indicated θ. Do *not* find θ.

308. $\sin\theta = \dfrac{3}{5}, \quad \cos\theta < 0$

309. $\tan\theta - \dfrac{4}{3}, \quad \sin\theta < 0$

Inverse Trigonometric Functions

For questions 310–312, find the value of the given expression. Do *not* use a calculator or a table of values.

310. $\sin^{-1}\left(\dfrac{1}{2}\right)$

311. $\arctan \sqrt{3}$

312. $\tan(\arccos 0.5)$

Graphing the Trigonometric Functions

For questions 313 and 314, find the period of the function.

313. $y = 4\sin(2x + \pi)$

314. $f(x) = 3 \tan (2x + \pi)$

For questions 315 and 316, determine the phase shift.

315. $y = \sin (x + \pi)$

316. $f(x) = \dfrac{1}{3} \sin \left(2x - \dfrac{\pi}{7} \right)$

For questions 317 and 318, sketch, on the same axes, one complete period of each function.

317. $y = \sin x$, $y = \sin 2x$

318. $y = \sin x$, $y = 2 \sin x$

For questions 319–321, sketch the graph of the given equation.

319. $y = \sin \left(x + \dfrac{\pi}{3} \right)$

320. $y = 3 \tan 2x \ (-\pi \leq x \leq \pi)$

321. $y = \arcsin x$

CHAPTER 8

Trigonometric Equations and Inequalities

Elementary Identities

For questions 322–330, prove that the given equation is an identity. Refer to Figure 8.1 when necessary.

Fundamental Relations	
$\sin\theta\csc\theta = 1$	$\tan\theta = \dfrac{\sin\theta}{\cos\theta}$
$\cos\theta\sec\theta = 1$	
$\tan\theta\cot\theta = 1$	$\cot\theta = \dfrac{\cos\theta}{\sin\theta}$
$\sin^2\theta + \cos^2\theta = 1$	
$1 + \cot^2\theta = \csc^2\theta$	
$1 + \tan^2\theta = \sec^2\theta$	

Figure 8.1

322. $\sin x(\csc x - \sin x) = \cos^2 x$

323. $\cos x(\sec x - \cos x) = \sin^2 x$

324. $\dfrac{\sin x + \cos x}{\tan x} = \cos x + \dfrac{\cos^2 x}{\sin x}$

325. $\dfrac{\sin^2 x}{1-\sin^2 x} + \dfrac{\cos^2 x}{1-\cos^2 x} = \dfrac{\tan^4 x + 1}{\tan^2 x}$

326. $(1 - \sin x)(1 + \sin x) + \sin^2 x = 1$

327. $\dfrac{1+\cot x}{\cot x} = \tan x + \csc^2 x - \cot^2 x$

328. $\dfrac{1+\cot x}{\cot x} = \tan x + \csc^2 x - \cot^2 x$

329. $(\sin x + \cos x)^4 = 1 + 4\sin x \cos x + 4(\sin x \cos x)^2$

330. $\dfrac{\cos^2 \theta - \sin^2 \theta}{\sin\theta \cos\theta} = \cot\theta - \tan\theta$

331. Express each of the other functions of θ in terms of $\sin \theta$.

332. Express each of the other functions of θ in terms of $\tan \theta$.

333. Using the fundamental relations, find the values of the functions of θ, given $\sin \theta = \frac{3}{5}$.

334. Using the fundamental relations, find the values of the functions of θ, given $\tan \theta = -\frac{5}{12}$.

335. Perform the indicated operations.
  (A) $(\sin \theta - \cos \theta)(\sin \theta + \cos \theta)$
  (B) $(\sin A + \cos A)^2$
  (C) $(\sin x + \cos y)(\sin y - \cos x)$
  (D) $(\tan^2 A - \cot A)^2$
  (E) $1 + \dfrac{\cos\theta}{\sin\theta}$
  (F) $1 - \dfrac{\sin\theta}{\cos\theta} + \dfrac{2}{\cos^2 \theta}$

336. Factor.
  (A) $\sin^2 \theta - \sin \theta \cos \theta$
  (B) $\sin^2 \theta + \sin^2 \theta \cos^2 \theta$
  (C) $\sin^2 \theta + \sin \theta \sec \theta - 6 \sec^2 \theta$
  (D) $\sin^3 \theta \cos^2 \theta - \sin^2 \theta \cos^3 \theta + \sin \theta \cos^2 \theta$
  (E) $\sin^4 \theta - \cos^4 \theta$

337. Simplify each of the following.
  (A) $\sec \theta - \sec \theta \sin^2 \theta$
  (B) $\sin \theta \sec \theta \cot \theta$
  (C) $\sin^2 \theta (1 + \cot^2 \theta)$
  (D) $\sin^2 \theta \sec^2 \theta - \sec^2 \theta$
  (E) $(\sin \theta + \cos \theta)^2 + (\sin \theta - \cos \theta)^2$
  (F) $\tan^2 \theta \cos^2 \theta + \cot^2 \theta \sin^2 \theta$

Addition and Subtraction Identities

Refer to Figure 8.2 when necessary for questions in this section.

$$\sin(\theta \pm \alpha) = \sin\theta\cos\alpha \pm \cos\theta\sin\alpha$$
$$\cos(\theta \pm \alpha) = \cos\theta\cos\alpha \mp \sin\theta\sin\alpha$$
$$\tan(\theta \pm \alpha) = \frac{\tan\theta \pm \tan\alpha}{1 \mp \tan\theta\tan\alpha}$$

Addition and Subtraction Identities

Figure 8.2

For **questions 338–340**, find the value of the given expression. Do *not* use tables or a calculator.

338. $\tan 75°$

339. $\tan\dfrac{7\pi}{12}$

340. $\dfrac{\tan 80° + \tan 40°}{1 - \tan 80° \tan 40°}$

341. Prove: $\dfrac{\sin(\theta + h) - \sin\theta}{h} = (\cos\theta)\dfrac{\sin h}{h} - (\sin\theta)\dfrac{1 - \cos h}{h}.$

342. Evaluate: $\sin\left[\cos^{-1}\left(-\dfrac{4}{5}\right) + \sin^{-1}\left(-\dfrac{3}{5}\right)\right]$

Double- and Half-Angle Identities

For **questions 343 and 344**, prove that the given formula is true for all *x*. Use Figure 8.3 when necessary.

Double- and Half-Angle Identities
$$\sin 2x = 2\sin x\cos x$$
$$\cos 2x = \cos^2 x - \sin^2 x = 1 - 2\sin^2 x = 2\cos^2 x - 1$$
$$\tan 2x = \frac{2}{\cot x - \tan x}$$
$$\sin\frac{x}{2} = \pm\sqrt{\frac{1 - \cos x}{2}}$$
$$\cos\frac{x}{2} = \pm\sqrt{\frac{1 + \cos x}{2}}$$
$$\tan\frac{x}{2} = \frac{\sin x}{1 + \cos x} = \frac{1 - \cos x}{\sin x}$$

Figure 8.3

343. $\sin 2x = 2 \sin x \cos x$

344. $\cos 2x = \cos^2 x - \sin^2 x = 1 - 2 \sin^2 x = 2 \cos^2 x - 1$

For questions 345 and 346, find the exact value of the given expression without using a table or calculator.

345. $\sin 22.5°$

346. $\tan 165°$

For questions 347–349, find the exact value without using a table or calculator.

347. $\sin\left(2 \cos^{-1} \dfrac{3}{5}\right)$

348. $\tan\left[2 \cos^{-1}\left(-\dfrac{4}{5}\right)\right]$

349. $\sin\left(\dfrac{x}{2}\right)$ if $\cos x = \dfrac{1}{3}, 0° < x < 90°$

Product and Sum Identities

350. State the four basic product identities.

351. State the four basic sum identities.

352. Express each of the following as a sum or difference:
  (A) $\sin 40° \cos 30°$
  (B) $\cos 110° \sin 55°$
  (C) $\cos 50° \cos 35°$
  (D) $\sin 55° \sin 40°$

353. Rewrite as a product.
  $\sin 20° + \sin 15°$

354. Express each of the following as a product.
  (A) $\sin 50° + \sin 40°$
  (B) $\sin 70° - \sin 20°$
  (C) $\cos 55° + \cos 25°$
  (D) $\cos 35° - \cos 75°$

355. Prove: $\dfrac{\sin 4A + \sin 2A}{\cos 4A + \cos 2A} = \tan 3A.$

Trigonometric Equations

For questions 356 and 357, determine whether the given number is a solution of the given equation.

356. $\dfrac{3\pi}{4}$, $1 + \tan\ x = 0$

357. $\dfrac{\pi}{2}$, $1 + \tan\ x = 0$

For questions 358 and 359, solve the given equation, finding all solutions which lie in the interval $(0, 2\pi)$.

358. $4 \cos^2 x - 3 = 0$

359. $2 \sin^2 x - 1 = 0$

For questions 360–362, solve for all x such that $0 \le x < 2\pi$.

360. $2 \sin x - \csc x = 1$

361. $\cos x\ -\sqrt{3} \sin x = 1$

362. $2 \cos^2 \dfrac{1}{2} x = \cos^2 x$

363. Solve: $\arccos 2x = \arcsin x$

364. Solve: $\arctan x + \arctan (1 - x) = \arctan \frac{4}{3}$

365. Solve: $\sec x + \tan x = 1$, $(0, 2\pi)$.

366. Solve for all real x, using a calculator. Give answer to four significant digits.

$\sin x = 0.2977$

CHAPTER **9**

Additional Topics in Trigonometry

Right Triangles

For questions 367–369, find the remaining parts of $\triangle ABC$, where $\angle C = 90°$. Round all angles to the nearest minute and all lengths to the nearest hundredth.

367. $b = 12$, $c = 13$

368. $\angle B = 17°50'$, $c = 3.45$

369. $b = 10$, $c = 12.6$

370. Find the perimeter of a hexagon inscribed in a circle of radius 5 m (assume a regular polygon).

371. If a train climbs at a constant angle of $1°23'$, how many vertical feet has it climbed after going 1 mi?

372. Find the diameter of the moon (to the nearest mile) if at 239,000 mi from earth it subtends an angle of $32'$ relative to an observer on the earth.

373. An object 4 ft tall casts a 3-ft shadow when the angle of elevation of the sun is θ. Find θ to the nearest degree.

374. If A is an acute angle:
  (A) Why is $\sin A < 1$?
  (B) When is $\sin A = \cos A$?
  (C) Why is $\sin A < \csc A$?
  (D) Why is $\sin A < \tan A$?
  (E) When is $\sin A < \cos A$?
  (F) When is $\tan A > 1$?

375. Find the values of the trigonometric functions of $45°$.

376. Find the values of the trigonometric functions of $30°$ and $60°$.

377. When the sun is $20°$ above the horizon, how long is the shadow cast by a building 150 ft high?

378. A tree 100 ft tall casts a shadow 120 ft long. Find the angle of elevation of the sun.

379. Find the length of the chord of a circle of radius 20 m subtended by a central angle of $150°$.

380. Find the height of a tree if the angle of elevation of its top changes from $20°$ to $40°$ as the observer advances 75 ft toward its base. See Figure 9.1.

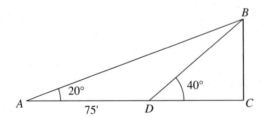

Figure 9.1

Law of Sines

For questions 381–384, find the measure of the indicated angle or the length of the indicated side. Refer to Figure 9.2, and use a calculator.

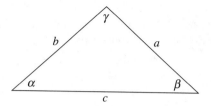

Figure 9.2

381. $\beta = 12°40'$, $\gamma = 100°$, $b = 13.1$; find a.

382. $\beta = 27°30'$, $\gamma = 54°30'$, $a = 9.27$; find b and c.

383. $\alpha = 50$, $c = 40$, $\gamma = 30°$; find b if $\alpha < 90°$.

384. $a = 14$, $b = 23$, $\alpha = 41°$; find β.

Law of Cosines

For questions 385–387, find the indicated piece of information concerning the triangle in Figure 9.2. Use a calculator.

385. $\alpha = 50°40'$, $b = 7.03$, $c = 7.00$; find a and β.

386. $\gamma = 120°20'$, $a = 5.73$, $b = 10.2$; find c and β.

387. $a = 4.00$, $b = 10.0$, $c = 9.00$; find α and γ.

Vectors

For questions 388 and 389, the magnitude and direction of a vector are given. Represent them geometrically.

388. 3, N40°E

389. 5, S60°W

For questions 390 and 391, find the magnitude of the given vector.

390. $(4, 5)$

391. $(6, -14)$

For questions 392 and 393, determine the direction of the given vector.

392. $(4, -3)$

393. $v = (-2, -1)$

394. Let $a = (1, 0)$, $b = (3, 0)$, $c = (4, 6)$, $d = (4, 9)$, $e = (1, 6)$. Find each of the following.
   (A) $2a$
   (B) $\dfrac{2}{3}b$
   (C) $a + c$
   (D) $a - d$
   (E) $a \cdot c$

395. Rewrite the given vector in terms of i and j.
   $(4, 7)$

396. Find the unit vector in the direction of the given vector.
   $(6, 4)$

397. Find the angle between the vectors $(1, 0)$ and $(\sqrt{2}, \sqrt{2})$.

398. An automobile weighing 2000 lb is standing on a smooth driveway that is inclined 5.0° with the horizontal. Find the force parallel to the driveway necessary to keep the car from rolling down the hill. Neglect all friction.

399. An airplane is moving horizontally at 240 mi/h when a bullet is shot from the plane with speed 2750 ft/s at right angles to the path of the airplane. Find the resultant speed and direction of the bullet.

400. A river flows due south at 125 ft/min. A motorboat, moving at 475 ft/min in still water, is headed due east across the river. (A) Find the direction in which the boat moves and its speed. (B) In what direction must the boat be headed in order for it to move due east, and what is its speed in that direction?

401. A block weighing $W = 500$ lb rests upon a ramp inclined 29° with the horizontal. Find the force tending to move the block down the ramp and the force of the block on the ramp.

CHAPTER 10

Conic Sections

The Circle

For questions 402–404, find the center and radius of the given circle.

402. $x^2 + y^2 = 100$

403. $7x^2 + 7y^2 + 14x - 56y - 25 = 0$

404. $x^2 + y^2 - 6x + 8y - 11 = 0$

405. Write the equation of a circle with center $(-2, -3)$ and radius $\sqrt{7}$.

406. Find the center and radius of the circle passing through the given points.

$(0, 0), (1, 1), (1, 2)$

The Parabola

For questions 407 and 408, find the focus and directrix for the given parabola.

407. $y^2 = 40x$

408. $x^2 = -16y$

For questions 409 and 410, sketch the graph to find the focus and directrix of the parabola given.

409. $y^2 = 8x$

410. $x^2 = 10y$

For questions **411** and **412**, find the vertex of the given parabola.

411. $x^2 - 6x + 8y + 25 = 0$

412. $y^2 - 16x + 2y + 49 = 0$

The Ellipse

For questions **413** and **414**, find the foci and lengths of the major and minor axes for each given ellipse.

413. $\dfrac{x^2}{25} + \dfrac{y^2}{4} = 1$

414. $\dfrac{x^2}{4} + \dfrac{y^2}{25} = 1$

For questions **415** and **416**, sketch the graph of the given ellipse.

415. $\dfrac{x^2}{25} + \dfrac{y^2}{4} = 1$

416. $\dfrac{x^2}{4} + \dfrac{y^2}{25} = 1$

417. Find the coordinates of the center of the ellipse and put the equation in standard form for an ellipse.
$x^2 + 4y^2 - 6x + 32y + 69 = 0$

418. Find the vertices and foci for the given ellipse.
$x^2 + 4y^2 - 6x + 32y + 69 = 0$

The Hyperbola

For questions **419–422**, find the foci and lengths of the transverse and conjugate axes.

419. $\dfrac{x^2}{9} - \dfrac{y^2}{4} = 1$

420. $\dfrac{y^2}{4} - \dfrac{x^2}{9} = 1$

421. $4x^2 - y^2 = 16$

422. $3y^2 - 2x^2 = 24$

For questions 423 and 424, sketch the given hyperbola. *Sketching hint*: Sketch first the rectangle formed by the two axes, then the diagonals to get the asymptotes, then the vertices and the hyperbola.

423. $\dfrac{x^2}{9} - \dfrac{y^2}{4} = 1$

424. $3y^2 - 2x^2 = 24$

For questions 425 and 426, find the equations of the asymptotes for the given hyperbola.

425. $x^2 - \dfrac{y^2}{4} = 1$

426. $\dfrac{y^2}{3} - \dfrac{x^2}{2} = 1$

For questions 427 and 428, find the center of the given hyperbola and put the equation in standard hyperbola form.

427. $x^2 - 4y^2 + 6x + 16y - 11 = 0$

428. $144x^2 - 25y^2 - 576x + 200y + 3776 = 0$

For questions 429 and 430, find the vertices, foci, and asymptotes.

429. $x^2 - 4y^2 + 6x + 16y - 11 = 0$

430. $144x^2 - 25y^2 - 576x + 200y + 3776 = 0$

Miscellaneous Problems

For questions 431 and 432, describe the locus represented by the given equation.

431. $x^2 + y^2 - 10x + 8y + 5 = 0$

432. $x^2 + y^2 + 4x - 6y + 24 = 0$

For questions 433–435, for the given parabola, sketch the curve, find the coordinates of the vertex and focus, and give the equations of the axis and directrix.

433. $y^2 = 16x$

434. $x^2 = -9y$

435. $x^2 - 2x - 12y + 25 = 0$

436. Write the equation of the conjugate of the hyperbola $25x^2 - 16y^2 = 400$, and sketch both curves.

CHAPTER 11

The Complex Numbers

Polar Form

437. Plot the given complex number.

    (A) $2 + i$

    (B) $2 + 3i$

    (C) $2 - i$

    (D) i

    (E) 6

For questions 438–440, give the polar (or trigonometric) form for the given complex number.

438. $4 + 0i$

439. $2\sqrt{3} - 2i$

440. $5 + 5i$

441. Perform the indicated operations.

$$\sqrt{2}\left[\cos\left(\frac{\pi}{4}\right) + i\sin\left(\frac{\pi}{4}\right)\right] \cdot 3\left[\cos\left(\frac{\pi}{2}\right) + i\sin\left(\frac{\pi}{2}\right)\right]$$

For questions 442–444, change the given complex number to polar form.

442. $\sqrt{2}\left[\cos\left(\frac{5\pi}{4}\right) + i\sin\left(\frac{5\pi}{4}\right)\right]$

443. $4(\cos 0 + i\sin 0)$

444. $4\left[\cos\left(\frac{11\pi}{6}\right) + i\sin\left(\frac{\pi}{6}\right)\right]$

For questions 445 and 446, perform the operation indicated, and write the result in $a + bi$ form.

445. $2\left[\cos\left(\dfrac{\pi}{6}\right) + i\sin\left(\dfrac{\pi}{6}\right)\right] \cdot 3\left[\cos\left(\dfrac{\pi}{3}\right) + i\sin\left(\dfrac{\pi}{3}\right)\right]$

446. $\dfrac{12(\cos 200° + i\sin 200°)}{3(\cos 350° + i\sin 350°)}$

447. Use polar form to perform the given calculation.
$(1+i)(\sqrt{2} - i\sqrt{2})$

Roots and De Moivre's Theorem

448. State De Moivre's theorem.

449. Find the indicated power.
$[3(\cos 10° + i\sin 10°)]^3$

For questions 450 and 451, use De Moivre's theorem to rewrite each in standard form.

450. $\left[\cos\left(\dfrac{\pi}{3}\right) + i\sin\left(\dfrac{\pi}{3}\right)\right]^5$

451. $(\sqrt{3} + i)^{12}$

For questions 452–454, find the indicated roots.

452. The square roots of i

453. The cube roots of -8

454. The square roots of $1 + i\sqrt{3}$

455. Solve: $x^3 + 1 = 0$

456. Solve: $x^2 - (1 + i\sqrt{3}) = 0$

CHAPTER 12

Sequences, Series, and Probability

Sequences

For questions 457–459, write the first four terms of the sequence whose general term is given.

457. $n - 3$

458. $\dfrac{1}{4^{n-2} + 1}$

459. $\dfrac{(-1)^n}{n+2}$

460. Find the general term.

$$\frac{1}{2}, \frac{3}{4}, \frac{5}{6}, \frac{7}{8}, \ldots$$

461. Find the first four terms of the sequence defined recursively by the given equations.

$$A_1 = 1, A_2 = 1, A_{n+2} = A_n + A_{n+1}$$

462. Define the given sequence by using a recursive formula.

$$1, 1, 5, 17, 61, \ldots$$

Series

463. Expand $\displaystyle\sum_{i=1}^{3} (-1)^{i+1}(x^i - 1)$

464. Write the given series, using summation (sigma) notation.

$$1 - 3 + 5 - 7 + 9 - 11$$

Arithmetic and Geometric Sequences

For questions 465 and 466, tell whether the given sequence is arithmetic or geometric.

465. $1, \dfrac{1}{2}, \dfrac{1}{4}, \dfrac{1}{8}, \ldots$

466. $10, 6, 2, -2, -6, \ldots$

467. Find all indicated quantities, where $a_1, a_2, \ldots, a_n, \ldots$ is an arithmetic sequence, $d =$ common difference, and $S =$ sum of first n terms of the sequence.

$a_1 = 1, a_2 = 5, S_{21} = ?$

468. Find the sum of the first eight terms of the sequence $1, \frac{1}{3}, \frac{1}{9}, \frac{1}{27}, \ldots$, the sequence is geometric, and r stands for the common ratio.

Geometric Series

469. Write the given geometric series, using sigma (summation) notation.

$$2 + 1 + \dfrac{1}{2} + \dfrac{1}{4} + \dfrac{1}{8} + \cdots$$

For questions 470–473, find the sum or show that the series has no sum.

470. $1 + \dfrac{1}{2} + \dfrac{1}{4} + \dfrac{1}{8} + \cdots$

471. $1 + \dfrac{1}{9} + \dfrac{1}{81} + \cdots$

472. $3 + 6 + 12 + 24 + \cdots$

473. $3 - \dfrac{3}{2} + \dfrac{3}{4} - \dfrac{3}{8} + \cdots$

Binomial Theorem

For questions 474–476, expand the given expression, using the binomial theorem.

474. $(x + y)^2$

475. $(2a + 3)^5$

476. $(2x - 4)^3$

Permutations, Combinations, and Probability

For questions 477–479, calculate the given permutation or combination.

477. $P(7, 1)$

478. $P(4, 1) + P(4, 2) + P(4, 3) + P(4, 4)$

479. $C(31, 2)$

480. How many three-digit numbers can be formed with the digits 1, 2, 3, 4, 5
    (A) if repetitions are allowed
    (B) if no repetitions are allowed

481. How many committees, consisting of 1 first-year student, 1 sophomore, and 1 junior can be selected from 40 first-year students, 30 sophomores, and 25 juniors?

482. Five boys are in a room that has 4 doors. In how many ways can they leave the room?

483. How many even numbers of three digits can be made with the digits 1, 2, 3, 4, 5, 6, 7 if no digit is repeated?

484. Seven songs are to be given in a program. In how many different orders could they be rendered?

485. Find the number of permutations that can be formed by using all the given letters.
ALABAMA

486. A line is formed by 5 girls and 5 boys, with boys and girls alternating. Find the number of ways of making the line.

487. Do as in question 486, but with a circular arrangement.

488. Show that $P(n, n - 1) = P(n, n)$.

489. In how many different ways can a tennis team of 4 be chosen from 17 players?

490. Nine points, no three of which are on a straight line, are marked on a chalkboard. How many lines, each through two of the points, can be drawn?

491. Seven different coins are tossed simultaneously. In how many ways can 3 heads and 4 tails come up?

492. In how many ways can a court of 9 judges make a 5-to-4 decision?

493. Four delegates are to be chosen from 8 members of a club.
   (A) How many choices are possible?
   (B) How many choices contain member A?
   (C) How many choices contain A and B?

494. The alphabet consists of 21 consonants and 5 vowels. In how many ways can 5 consonants and 3 vowels be selected?

495. A ball is chosen from a bag containing 2 white, 1 black, and 2 red balls.
   (A) Find the probability that a white ball is chosen.
   (B) Find the probability that a black ball is chosen.
   (C) Find the probability of choosing a green ball.

496. Three pennies are tossed at the same time.
   (A) Find the probability that all 3 land on heads (H).
   (B) Find the probability that 2 are H and 1 is T (tails).
   (C) Find the probability that 1 is H and 2 are T.

497. Three balls are drawn from a bag containing 6 red and 5 black balls.
   (A) Find the probability that all are red.
   (B) Find the probability that all are black.
   (C) Find the probability of 2 being red and 1 being black.

498. One card is picked from an ordinary 52-card deck.
   (A) Find the probability that a red card is chosen.
   (B) Find the probability that a queen is chosen.
   (C) Find the probability that a red 8 is chosen.

499. The probability that a brush salesperson will make a sale at one house is $\frac{2}{3}$ and at a second house is $\frac{1}{2}$.

(A) Find the probability the salesperson will make both sales.

(B) Find the probability he or she will make neither sale.

(C) Find the probability she or he will make at least one sale.

(D) Find the probability the salesperson will make *exactly* one sale.

500. Bag 1 contains 2 white balls and 1 red ball; bag 2 contains 3 white balls and 1 red ball. A bag is chosen at random, and a ball is picked at random. What is the given probability?

(A) The ball chosen is red.

(B) The ball chosen is white.

ANSWERS

Chapter 1: Review of Basic Algebra

1. (A) For any real number a, $a + (-a) = 0$. We say $-a$ is the *additive identity* of a. Here replace a by -1 and $a + (-a) = 0 = (-1) + [-(-1)]$.

(B) If t and u are real and $u \neq 0$, then $t \div u = t \left(\dfrac{1}{u} \right)$ by the definition of division. In this case, $t = 7$, $u = 9$ (and thus $u \neq 0$), and we have a direct application of the definition of division.

(C) For any real number r, $1 \cdot r = r$. Here 1 is the *multiplicative identity*. In this case, r is simply replaced by $-\frac{2}{3}$.

(D) If a and b are real numbers, then $a - b$ is defined as $a + (-b)$. Thus, $8 - 12 = 8 + (-12)$ is an application of the definition of *subtraction*.

(E) $a(b + c) = ab + ac$ is the *distributive law* for real numbers. Here $a = 7$, $b = s$, $c = t$.

(F) $(a + b) + c = a + (b + c)$ is the *associative law* for addition of real numbers. In this case, $a = 3p$, $b = 9$, $c = 3$.

(G) Here ym is being replaced by my. And $ym = my$ is the *commutative law* for multiplication of real numbers.

2. (A) Recall that, for any two signed numbers of like sign, the product is nonnegative; for any two of unlike signs, the product is nonpositive. Thus

$$(-2)(3) = -6 \text{ (unlike signs)}$$
$$(-6)(-5) = +30 \text{ (like signs)}$$

Therefore,
$$(-2)\underbrace{(3)(-5)} = (-6)(-5)$$
$$= +30$$

(B) To add two signed numbers of unlike sign, we subtract the smaller (in absolute value) from the larger and give the result the sign of the larger. Here 40 is larger than 7. Thus, we find $40 - 7 = 33$, and we use the sign of 40 (which is +). Thus, the answer is 33.

(C) Remove parentheses first: $-(-6) = 6$. Then

$$-8 + 6 + 2 = (-8 + 6) + 2$$
$$= -2 + 2$$
$$= 0$$

(D) First simplify the numerator and denominator separately; then perform the division. (By the way, this is the first of several times you will be seeing complex fractions in this book.)

$$3 - \frac{2}{3} = \frac{9}{3} - \frac{2}{3} = \frac{7}{3};$$

$$5 + \frac{5}{6} = \frac{30}{6} + \frac{5}{6} = \frac{35}{6}.$$

$$\frac{\frac{7}{3}}{\frac{35}{6}} = \frac{7}{1} \cdot \frac{2}{35} = \frac{14}{35} = \frac{2}{5}.$$

3. (A) Recall that $(a + bi) \pm (c + di) = (a \pm c) + (b \pm d)i$. Then $(1 + i) + (3 - 2i) = 4 + (-1)i = 4 - i$.

(B) $(2 - i) - (3 - 4i) = (2 - 3) + (-1 + 4)i = -1 + 3i$.

(C) Recall that $(a + bi)(c + di) = (ac - bd) + (bc + ad)i$. Then $(2 + i)(3 + 2i) = 2(3) - 1(2) + (3 + 4)i = 4 + 7i$.

(D) $\dfrac{2+i}{1-i} = \dfrac{2+i}{1-i} \cdot \dfrac{1+i}{1+i} = \dfrac{1+3i}{1+1} = \dfrac{1+3i}{2} = \dfrac{1}{2} + \dfrac{3}{2}i$.

(E) $(2 + \sqrt{-9})(1 + \sqrt{-4}) = (2 + 3i)(1 + 2i) = -4 + 7i$.

4. $i^{4k} = (i^4)^k = 1^k = 1$.

5. Two complex numbers $a + bi$ and $c + di$ are equal if and only if $a = c$ and $b = d$. So $3 - 2i = 4xi + 2y$ is rewritten as $3 - 2i = 2y + 4xi$; then $2y = 3$ and $4x = -2$, or $y = \frac{3}{2}$ and $x = -\frac{1}{2}$.

6. (A) $\dfrac{7^6}{7^4} = 7^{6-4} = 7^2 = 49$. $\left(Recall: \dfrac{x^a}{x^b} = x^{a-b}.\right)$

(B) Recall that $x^a \cdot x^b = x^{a+b}$. Thus, $3^{41} \cdot 3^{-9} = 3^{41+(-9)} = 3^{32}$.

(C) $\dfrac{(2+2^{-1})}{5} + (-8)^0 - 4^{\frac{3}{2}} = \dfrac{(2+\frac{1}{2})}{5} + 1 - (4^{\frac{1}{2}})^3 = \dfrac{\frac{5}{2}}{5} + 1 - 2^3 = \dfrac{5}{10} + 1 - 8 = \dfrac{1}{2} - 7 = -6\dfrac{1}{2}$.

(D) $125^{-\frac{4}{3}} = (125^{\frac{1}{3}})^{-4} = 5^{-4} = \dfrac{1}{5^4} = \dfrac{1}{625}$.

7. $(2ab^2)^3 (a^2c)^2 = (2^3a^3b^6)(a^4c^2) = (8a^3b^6)(a^4c^2) = 8a^7b^6c^2$.

8. $\dfrac{x^{-8} \cdot x^{-7}}{x^{-6}} \div \dfrac{x^{-5} \cdot x^{-4}}{x^{-3}} = \dfrac{x^{-15}}{x^{-6}} \cdot \dfrac{x^{-3}}{x^{-5} \cdot x^{-4}} = \dfrac{x^{-15}}{x^{-6}} \cdot \dfrac{x^{-3}}{x^{-9}} = x^{-9} \cdot x^6 = x^{-3} = \dfrac{1}{x^3}$.

9. $\dfrac{\sqrt{a} \cdot a^{-\frac{2}{3}}}{\sqrt[6]{a^5}} + \dfrac{a^{-\frac{5}{6}}}{\sqrt[3]{a^2} \cdot a^{-\frac{1}{2}}} = \dfrac{a^{\frac{1}{2}}a^{-\frac{2}{3}}}{a^{\frac{5}{6}}} + \dfrac{a^{-\frac{5}{6}}}{a^{\frac{2}{3}}a^{-\frac{1}{2}}}$

$$= \dfrac{a^{-\frac{1}{6}}}{a^{\frac{5}{6}}} + \dfrac{a^{-\frac{5}{6}}}{a^{\frac{1}{6}}} = \dfrac{a^{-\frac{1}{6}} + a^{\frac{4}{6}}a^{-\frac{5}{6}}}{a^{\frac{5}{6}}}$$

$$= \dfrac{a^{-\frac{1}{6}} + a^{-\frac{1}{6}}}{a^{\frac{5}{6}}} = \dfrac{2a^{-\frac{1}{6}}}{a^{\frac{5}{6}}}$$

$$= \dfrac{2}{a^{\frac{1}{6}}a^{\frac{5}{6}}} = \dfrac{2}{a^{\frac{6}{6}}} = \dfrac{2}{a}.$$

10. $2a\sqrt[3]{8a^8b^{13}} = 2a\sqrt[3]{2^3(a^2)^3a^2(b^4)^3b} = 2a\sqrt[3]{2^3}\sqrt[3]{(a^2)^3a^2}\sqrt[3]{(b^4)^3b} = 2a\cdot2\cdot a^2\cdot\sqrt[3]{a^2}\cdot b^4\cdot\sqrt[3]{b} = 4a^3b^4\cdot\sqrt[3]{a^2b}.$

11. $\sqrt[3]{\dfrac{3y^5}{4x^4}} = \sqrt[3]{\dfrac{3y^3y^2}{4x^3x}} = \dfrac{\sqrt[3]{3y^3y^2}}{\sqrt[3]{4x^3x}} = \dfrac{y\sqrt[3]{3y^2}}{x\sqrt[3]{4x}}\cdot\dfrac{\sqrt[3]{4^2x^2}}{\sqrt[3]{4^2x^2}}$ (rationalizing the denominator)

$= \dfrac{y\sqrt[3]{48x^2y^2}}{x\sqrt[3]{4^3x^3}} = \dfrac{y\sqrt[3]{48x^2y^2}}{x\cdot4\cdot x} = \dfrac{y\sqrt[3]{48x^2y^2}}{4x^2}.$

12. $\dfrac{2}{x^2 - \sqrt{x^4 + 2x^2 + 1}} = \dfrac{2}{x^2 - \sqrt{(x^2+1)^2}} = \dfrac{2}{x^2 - (x^2+1)} = \dfrac{2}{x^2 - x^2 - 1} = -2.$

13. $(x^2 - 2x + 3) + (4x^2 - x + 6) = x^2 - 2x + 3 + 4x^2 - x + 6 = (x^2 + 4x^2) + (-2x - x) + 3 + 6 = 5x^2 - 3x + 9.$

14. $x^2 - 2x + 3 - (-5x^3 - 7x + 1) = x^2 - 2x + 3 + 5x^3 + 7x - 1 = 5x^3 + x^2 + (7x - 2x) + 3 - 1 = 5x^3 + x^2 + 5x + 2.$

15. $(x - 2y)(x + 3y) = x(x) + x(3y) + (-2y)(x) + (-2y)(3y) = x^2 + 3xy - 2xy - 6y^2 = x^2 + xy - 6y^2.$

16. $(\sqrt{s} + \sqrt{t})^2 = (\sqrt{s})^2 + 2\sqrt{s}\sqrt{t} + (\sqrt{t})^2 = s + 2\sqrt{st} + t.$

17. $(s + t)^3 = (s + t)^2(s + t) = (s^2 + 2st + t^2)(s + t) = s^2(s) + 2st(s) + t^2(s) + s^2(t) + 2st(t) + t^2(t) = s^3 + 3s^2t + 3st^2 + t^3.$

18. $m - \{m - [m - (m - 1)]\} = m - \{m - [m - m + 1]\} = m - \{m - 1\} = m - m + 1 = 1.$

19. Substituting -5 for x gives $2x^2 - 3x - 10 = 2(-5)^2 - 3(-5) - 10 = 2(25) + 15 - 10 = 50 + 15 - 10 = 55.$

20. (A) Let $n =$ the number. Then "27 more than the number" means 27 more than n, which is 27 plus n. Thus, $27 + n$ is the solution.

(B) The number is n. Thus, the number increased by 45 is the number plus 45, which is $n + 45$.

(C) 43 less than 90 is $90 - 43$. Similarly, 43 less than the number is 43 less than $n = n - 43$.

(D) 43 less 10 means $43 - 10$; thus, 43 less the number means $43 - n$.

21. We are looking for two binomials $cx + d$ and $ex + f$ such that $x^2 + 4x + 3 = (cx + d)(ex + f)$. Note that $(cx + d)(ex + f) = cex^2 + (de + cf)x + df$. Thus, $x^2 + 4x + 3 = cex^2 + (de + cf)x + df$. In order for this to take place, ce must be equal to the coefficient of x^2 (in this case, 1), and df must be 3. We conclude that $de + cf = 4$, $c = e = 1$, and either $f = 3$, $d = 1$ or $f = 1$, $d = 3$. Thus, $f = 3$, $d = 1$ or $f = 1$, $d = 3$. So $x^2 + 4x + 3 = (x + 3)(x + 1).$

Note that $(x+1)(x+3)$ is also correct. We can check our answer by noting that $(x+3)(x+1) = x^2 + (3x+x) + 3 = x^2 + 4x + 3$.

22. We repeat the procedure given in Question 21. We want $ce = 2$, $de + cf = 7$, and $df = 6$. Then $c = 2$, $e = 1$ (or the reverse is fine also), $de + cf = 7$, and $d = 6$, $f = 1$ or $d = 3$, $f = 2$ or $d = 1$, $f = 6$ or $d = 2$, $f = 3$. Note that since $ce = 2$, $d = 6$, $f = 1$ is different from $d = 1$, $f = 6$. Checking these choices, we find that if $d = 3$ and $f = 2$, then $de + cf = 7$. So $2x^2 + 7x + 6 = (2x + 3)(x + 2)$.

23. Recall that a polynomial of the form $r^2 - s^2$ is factored as follows: $r^2 - s^2 = (r+s)(r-s)$ (the "difference of two perfect squares" formula). See Figure A1.1. In this case, we notice that $4y^2 = (2y)^2$ and $25 = 5^2$, and we conclude that $4y^2 - 25 = (2y + 5)(2y - 5)$. Check to see that $(2y + 5)(2y - 5) = 4y^2 - 25$ when you multiply.

(*a*) Difference of two squares:

$$r^2 - s^2 = (r+s)(r-s)$$

(*b*) Difference of two cubes:

$$r^3 - s^3 = (r-s)(r^2 + rs + s^2)$$

(*c*) Sum of two cubes:

$$(r^3 + s^3) = (r+s)(r^2 - rs + s^2)$$

Figure A1.1

24. We check the discriminant since we are having trouble finding factors. In this case, the discriminant $b^2 - 4ac$ is $(-6)^2 - 4(1)(-3) = 36 + 12 = 48$, which is not a perfect square. If the discriminant is not a perfect square, then the polynomial is not factorable relative to the integers.

25. Try to find factors here. You will discover that they are difficult to find. Check the discriminent $b^2 - 4ac$ (where $ax^2 + bx + c$ is the form of the polynomial). In this case, $b^2 - 4ac$ is $1^2 - 4(1)(1)$, which is negative. A negative discriminant signals that the polynomial is not factorable over the integers.

26. This polynomial is of the form of a sum of two cubes: $27p^3 + 8q^3 = (3p)^3 + (2q)^3$. Look at Figure A1.1 for the correct formula. Then $27p^3 + 8q^3 = (3p + 2q)[(3p)^2 - (3p)(2q) + (2q)^2] = (3p + 2q)(9p^2 - 6pq + 4q^2)$.

27. Notice that each term contains a factor of $2x^2$. Factor out that $2x^2$. $2x^4 - 24x^3 + 40x^2 = 2x^2(x^2 - 12x + 20)$. Now factor $x^2 - 12x + 20$. $x^2 - 12x + 20 = (x - 10)(x - 2)$. Thus, $2x^4 - 24x^3 + 40x^2 = 2x^2(x - 10)(x - 2)$.

28. We will use a regrouping technique here. $4x^2 - 9y^2 + 4x + 1 = 4x^2 + 4x + 1 - 9y^2 = (2x+1)$
$(2x+1) - 9y^2 = (2x+1)^2 - 9y^2$ (difference of two squares) $= (2x+1-3y)(2x+1+3y)$.

29. $a^4 - b^4 = (a^2)^2 - (b^2)^2 = (a^2 - b^2)(a^2 + b^2) = (a+b)(a-b)(a^2+b^2)$. Notice that we are applying the "difference of two squares" formula twice here.

30. $\dfrac{x^2 - 4}{x^2 - 4x - 12} = \dfrac{\cancel{(x+2)}(x-2)}{(x-6)\cancel{(x+2)}} = \dfrac{x-2}{x-6}$.

31. $\dfrac{4m-3}{18m^3} + \dfrac{3}{m} - \dfrac{2m-1}{6m^2} = \dfrac{(4m-3) + 3(18m^2) - (2m-1)(3m)}{18m^3}$

$$(18m^3 \text{ is the least common denominator})$$

$$= \frac{4m - 3 + 54m^2 - (6m^2 - 3m)}{18m^3} = \frac{4m - 3 + 54m^2 - 6m^2 + 3m}{18m^3}$$

$$= \frac{48m^2 + 7m - 3}{18m^3}.$$

32. $\dfrac{x^2}{12} + \dfrac{x}{18} - \dfrac{1}{30} = \dfrac{x^2(15) + x(10) - 1(6)}{180}$ (180 is the least common denominator) $=$
$\dfrac{15x^2 + 10x - 6}{180}$.

33. $\dfrac{x^2 - 2x - 15}{x^3 - 3x^2 - 10x} = \dfrac{x^2 - 2x - 15}{x(x^2 - 3x - 10)} = \dfrac{\cancel{(x-5)}(x+3)}{x\cancel{(x-5)}(x+2)} = \dfrac{x+3}{x(x+2)}$.

34. $\dfrac{2-x}{2x+x^2} \cdot \dfrac{x^2 + 4x + 4}{x^2 - 4} = \dfrac{2-x}{x(2+x)} \cdot \dfrac{(x+2)(x+2)}{(x+2)(x-2)}$

$$= \frac{-\cancel{(x-2)}}{x\cancel{(2+x)}} \cdot \frac{\cancel{(x+2)}\,\cancel{(x+2)}}{\cancel{(x+2)}\,\cancel{(x-2)}} = \frac{-1}{x}.$$

[Notice the $2 - x = -(x-2)$ trick here. Also remember that $2 + x$ and $x + 2$ are equal, and so they cancel!]

35. $\dfrac{2}{x^2 - 4} + \dfrac{6}{x-2} = \dfrac{2}{(x+2)(x-2)} + \dfrac{6}{x-2} = \dfrac{2(1) + 6(x+2)}{(x+2)(x-2)} = \dfrac{2 + 6x + 12}{(x+2)(x-2)}$

[$(x+2)(x-2)$ is the least common denominator] $= \dfrac{6x + 14}{x^2 - 4}$.

36. $\dfrac{1}{x-1} + \dfrac{2x}{(x-1)^2} - \dfrac{4-x}{(x-1)^3} = \dfrac{1(x-1)^2 + 2x(x-1) - 1(4-x)}{(x-1)^3}$

$$= \frac{(x^2 - 2x + 1) + (2x^2 - 2x) - (4 - x)}{(x-1)^3}$$

$$= \frac{3x^2 - 3x - 3}{(x-1)^3} = \frac{3(x^2 - x - 1)}{(x-1)^3}.$$

37. Notice that this expression involves a complex fraction. There are two basic ways of solving problems like this one. (1) Get rid of the complex fraction first, and then simplify.

$$\frac{2+\dfrac{1}{t}}{2-\dfrac{1}{t}} = \frac{\left(2+\dfrac{1}{t}\right)}{\left(2-\dfrac{1}{t}\right)} \cdot \frac{t}{t} = \frac{2t+1}{2t-1}$$

(t is the least common denominator of $\dfrac{1}{t}$ and $\dfrac{1}{t}$.)

(2) Simplify the numerator and denominator first:

$$\frac{2+\dfrac{1}{t}}{2-\dfrac{1}{t}} = \frac{\dfrac{(2t+1)}{t}}{\dfrac{(2t-1)}{t}} = \frac{2t+1}{t} \div \frac{2t-1}{t} = \frac{2t+1}{t} \cdot \frac{t}{2t-1} = \frac{2t+1}{2t-1}.$$

38. $\dfrac{x+y}{x^{-1}+y^{-1}} = \dfrac{x+y}{\dfrac{1}{x}+\dfrac{1}{y}} = \dfrac{x+y}{\dfrac{1}{x}+\dfrac{1}{y}} \cdot \dfrac{xy}{xy} = \dfrac{xy(x+y)}{y+x} = xy.$

39. $\dfrac{1-\sqrt{x+1}}{1+\sqrt{x+1}} = \dfrac{1-\sqrt{x+1}}{1+\sqrt{x+1}} \cdot \dfrac{1-\sqrt{x+1}}{1-\sqrt{x+1}}$

(Remember! $1-\sqrt{x+1}$ is the conjugate of $1+\sqrt{x+1}$. It is always the case that the conjugate of $\sqrt{m}+\sqrt{n}$ is $\sqrt{m}-\sqrt{n}$; the conjugate of $\sqrt{m}-\sqrt{n}$ is $\sqrt{m}+\sqrt{n}$.)

$$= \frac{1-2\sqrt{x+1}-(x+1)}{1-(x+1)} = \frac{1-2\sqrt{x+1}-x-1}{1-x-1} = \frac{-2\sqrt{x+1}-x}{-x}.$$

40. $\dfrac{\dfrac{\sqrt{2+h}+\sqrt{2}}{h}}{\dfrac{1}{\sqrt{2+h}-\sqrt{2}}} = \dfrac{\sqrt{2+h}+\sqrt{2}}{h} \cdot \dfrac{\sqrt{2+h}-\sqrt{2}}{\sqrt{2+h}-\sqrt{2}} = \dfrac{2+h-2}{h(\sqrt{2+h}-\sqrt{2})} = \dfrac{h}{h(\sqrt{2+h}-\sqrt{2})} =$

(Notice that we use the same procedure used for rationalizing the denominator.)

Chapter 2: Equations and Inequalities

41.

$$6x + 3 = 19x + 5$$

$$\underline{ -3 \qquad\qquad -3} \qquad \text{(Add } -3 \text{ to both sides.)}$$

$$6x = 19x + 2$$

$$\underline{-19x \qquad\qquad -19x} \qquad \text{(Add } -19x \text{ to both sides.)}$$

$$\underline{\left(-\frac{1}{13}\right)(-13x) = 2 \cdot \left(-\frac{1}{13}\right)} \qquad \left(\text{Multiply both side by } -\frac{1}{13}.\right)$$

$$x = -\frac{2}{13}$$

Note that we added −3 to both sides so that constant terms could be found on only one side of the equation [in this case, the right-hand side (RHS)]. We then add −19x to both sides to isolate the terms involving a variable. We multiplied by $-\frac{1}{13}$ to transform −13x to x.

42. First, remove parentheses: $6x - 12 + 12x - 20 = 14x$. Next, collect like terms on the left-hand side (LHS): $(12x + 6x) - 12 - 20 = 14x$.

$$18x - 32 = 14x$$

$$\underline{-18x \qquad\qquad -18x}$$

$$-32 = -4x$$

$$\frac{32}{4} = x$$

$$x = 8$$

43. $x - 2 + 6x = 6 + 12 - 3x$

$$7x - 2 = 18 - 3x$$

$$10x = 20$$

$$x = 2$$

Check:

$$2 - 2(1 - 6) \overset{?}{=} 6 + 3(4 - 2)$$

$$12 = 12$$

44. $ay - cy = d - b$

$$(a - c)y = d - b$$

$$y = \frac{d - b}{a - c}$$

Check:

$$a\left(\frac{d - b}{a - c}\right) + b \overset{?}{=} c\left(\frac{d - b}{a - c}\right) + d$$

$$\frac{ad - ab}{a - c} + b \overset{?}{=} \frac{cd - cb}{a - c} + d$$

$$\frac{ad - ab + b(a - c)}{a - c} \overset{?}{=} \frac{cd - cb + d(a - c)}{a - c}$$

$$\frac{ad - ab + ba - bc}{a - c} = \frac{cd - cb + da - dc}{a - c}$$

45. $6x - 4 = 40 - 5x$ (Multiplication by what?)

$$11x = 44$$
$$x = 4$$

Check:
$$\frac{3(4) - 2}{5} \stackrel{?}{=} 4 - \frac{1}{2}(4)$$
$$2 = 2$$

46. $(3x + 1)(2x - 3) = (2x + 1)(3x - 1)$

$$6x^2 - 7x - 3 = 6x^2 + x - 1$$
$$-8x = 2$$
$$x = -\frac{1}{4}$$

Check:
$$\frac{3\left(-\frac{1}{4}\right) + 1}{3\left(-\frac{1}{4}\right) - 1} \stackrel{?}{=} \frac{2\left(-\frac{1}{4}\right) + 1}{2\left(-\frac{1}{4}\right) - 3}$$
$$-\frac{1}{7} = -\frac{1}{7}$$

47. $x + 1 - (x - 3) = 3x - 2$

$$-3x = -6$$
$$x = 2$$

Check:
$$-1 - \frac{1}{3} \stackrel{?}{=} \frac{6 - 2}{-3}$$
$$-\frac{4}{3} = -\frac{4}{3}$$

48. $x^2 - 2 = (x + 1)(x - 1) - 1$

$$= x^2 - 2$$

Thus, the solution set is the set of real numbers except $x = 1$ (since division by 0 is not defined).

49. $x - 3 + (x - 1) = 2x - 5$

$$2x - 4 = 2x - 5$$

But there is no x such that $2x - 4 = 2x - 5$. The solution set is $\varnothing$.

50. (A) If $x = 12$ is a solution, then $2(12) + 5$ must equal $3(12) + k$. Then $24 + 5 = 36 + k$, $29 = 36 + k$, or $-7 = k$.
   (B) Substitute 2 for x. Then $2^2 + k(2) + 2 = 0$, $4 + 2k + 2 = 0$, $2k = -6$, or $k = -3$.

51. Let x = one of the integers; then $x + 2$ must be the other. (Consecutive even integers differ by 2.) Thus, $x + (x + 2) = 10$, $2x + 2 = 10$, $2x = 8$, or $x = 4$, then $x + 2 = 6$. The integers are 4 and 6. *Check*: 4 and 6 are consecutive even numbers. $4 + 6 = 10$. Thus, their sum is 10.

52. Let x = width. Then $2x$ = length (i.e., twice the width). Then $2x + 2x + x + x = 30$, $6x = 30$, $x = 5$ m (width), and $2x = 10$ m (length). *Check*: $10 = 2(5)$. Also $10 + 10 + 5 + 5 = 30$.

53. Let x = distance A travels until the trains meet. Then $60 - x$ = distance B travels. Put that information in Figure A2.1. Also insert the respective rates of the trains. Note that if $d = rt$, then $t = \frac{d}{r}$. Fill in these times. Notice that the two times are equal, since both trains must travel until they meet. So $\frac{x}{30} = \frac{(60 - x)}{45}$, $45x = 1800 - 30x$, $75x = 1800$, $x = 24$ mi = distance A travels.

	$d =$	r	$\times$	t
A	x	30		$\dfrac{x}{30}$
B	$60 - x$	45		$\dfrac{60 - x}{45}$

Figure A2.1

54. Let x = price before discount. Then $x - 20\% \, x$ = price after discount. $x - 0.2x = 72$, $0.8x = 72$, or $x = 72 \times \frac{10}{8} = \$90$.

55. Let x = time in hours it would take Kate's helper to paint the room. If it takes x h to complete, then $\frac{1}{x}$ of the job could be completed in 1 h. Also if Kate can complete the job in 6 h, she will complete $\frac{1}{6}$ of it in 1 h. Then in 1 h, $\frac{1}{6} + \frac{1}{x} = \frac{1}{\frac{24}{7}} \left(\frac{24}{7} = 3\frac{3}{7} \right), \frac{1}{6} + \frac{1}{x} = \frac{7}{24}, \frac{(x+6)}{(6x)} = \frac{7}{24}$, $24x + 144 = 42x$, $144 = 18x$, or $x = 8$ h. Her helper must be able to paint the room working alone in 8 h. *Check*: $\frac{1}{6} + \frac{1}{8} = \frac{7}{24} = \frac{1}{\frac{24}{7}}$.

56. Let x = Mary's age. Then $2x$ = Barbara's age, and $6x$ = Dick's age. $\frac{6x + 2x + x}{3} = 36$ (average of ages), $9x = 108$, so $x = 12$ years old, $2x = 24$ years old = Barbara's age, and $6x = 72$ years old = Dick's age.

57. Let x = number of nickels, $2x$ = number of dimes. Then, $5(x) + 10(2x) = 100(100\cent = \$1)$, $5x + 20x = 100$, $25x = 100$, or $x = 4$ nickels. *Check*: $2x = 8$ = number of dimes. $4(5\cent) + 8(10\cent) = 100\cent = \$1$.

58. If $x^2 = 9$, then x is that number or numbers which, when squared, is 9. Thus, $x = \pm 3$. Notice that, to get this result, we found the square roots (+ and −) of the constant term isolated on one side of the equation. Notice that there is no linear term in this equation. *Check*: $3^2 = 9$; $(-3)^2 = 9$.

59. $t^2 + 5 = 0$, or $t^2 = -5$. But no real number squared is negative. Thus, the solutions must be complex numbers. If $t^2 = -5$, then $t = \pm\sqrt{-5} = \pm i\sqrt{5}$. (see Chapter 11 for a review of complex numbers.)

60. $4x^2 - 7 = 0$, $4x^2 = 7$, $x^2 = \dfrac{7}{4}$, $x = \pm\sqrt{\dfrac{7}{4}} = \pm\dfrac{\sqrt{7}}{\sqrt{4}} = \pm\dfrac{\sqrt{7}}{2}$.

61. $(n+5)^2 = 9$, or $n+5 = \pm 3$. Possibility 1: $n + 5 = 3$, or $n = -2$. Possibility 2: $n + 5 = -3$, or $n = 8$. So the answer is $n = -2, -8$.

62. Factor the LHS: $(x + 2)(x + 1) = 0$ ($x + 2$ or $x + 1$ or both must be 0). If $x + 2 = 0$, then $x = -2$. If $x + 1 = 0$, then $x = -1$. *Check*: $(-2)^2 + 3(-2) + 2 = 4 - 6 + 2 = 0$, or $0 = 0$. Also $(-1)^2 + 3(-1) + 2 = 0$, $1 - 3 + 2 = 0$, or $0 = 0$.

63. $2t^2 - 2t = 12$, $t^2 - t = 6$, $t^2 - t - 6 = 0$, or $(t - 3)(t + 2) = 0$. If $t - 3 = 0$, then $t = 3$. If $t + 2 = 0$, then $t = -2$.

64. This does not appear to be factorable. Using the quadratic formula $x = \dfrac{-b \pm \sqrt{b^2 - 4ac}}{2a}$ ($a \neq 0$), we have $a = 1$, $b = 1$, $c = -1$, and $x = \dfrac{-1 \pm \sqrt{1^2 - 4(1)(-1)}}{2(1)} = \dfrac{-1 \pm \sqrt{5}}{2}$. Thus, $x = \dfrac{-1 \pm \sqrt{5}}{2}$ or $x = \dfrac{-1 - \sqrt{5}}{2}$.

65. Here we illustrate the method of "completing the square." $x^2 + 4x + 4$ is a perfect square, since $x^2 + 4x + ④ = (x + 2)^2$. To obtain the circled 4, we find the coefficient of the x term (here it is 4), divide it by 2, and square it: $(\frac{4}{2})^2 = 2^2 = 4$. $x^2 + 4x + 1 \neq x^2 + 4x + 4$, but $x^2 + 4x + 1 = (x^2 + 4x + 4) - 3$. So the equation $x^2 + 4x + 1 = 0$ is rewritten as $(x^2 + 4x + 4) - 3 = 0$, $(x + 2)^2 - 3 = 0$, $(x + 2)^2 = 3$, $x + 2 = \pm\sqrt{3}$, $x = \pm\sqrt{3} - 2$ or $x = \sqrt{3} - 2, -\sqrt{3} - 2$.

66. Either $(x^2 - 3)$ or $(x^2 - 4)$ (or both) $= 0$. So $x^2 - 3 = 0$, $x = \pm\sqrt{3}$, or $x^2 - 4 = 0$, $x = \pm 2$. Thus, $x = 3, -3, 2,$ or -2.

67. Let $u = \dfrac{1}{y} = y^{-1}$. Then $u^2 = \left(\dfrac{1}{y}\right)^2 = y^{-2}$, $8u^2 + 6u + 1 = 0$. $(4u + 1)(2u + 1) = 0$, or $u = -\frac{1}{4}, -\frac{1}{2}$. Then $\dfrac{1}{y} = -\frac{1}{4}$ or $-\frac{1}{2}$, or $y = -4, -2$.

68. **(A)** The discriminant of a quadratic equation $ax^2 + bx + c = 0$ is $b^2 - 4ac$. In this case, $b^2 - 4ac = 5^2 - 4(1)(-6) = 25 + 24 = 49$.
  (B) $a = 1$, $b = 5$, $c = -3$, $b^2 - 4ac = 25 - 4(1)(-3) = 25 + 12 = 37$.

69. **(A)** Recall that when the discriminant is positive or zero, the equation has real roots. If the discriminant is zero, it has one real root; otherwise, it has two. If the discriminant is negative, the equation has imaginary roots. Here, $b^2 - 4ac = 11^2 - 4(1)(11) > 0$. Conclusion: Two real roots.
  (B) $b^2 - 4ac = (-3)^2 - 4(1)\left(\frac{9}{4}\right) = 9 - 9 = 0$. Conclusion: One real root.
  (C) $b^2 - 4ac = 1^2 - 4(2)(1) < 0$. Conclusion: No real roots.

70. $\sqrt{3w-2} = 2 + \sqrt{w}$, $3w - 2 = (2 + \sqrt{w})^2$ (squaring), $3w - 2 = 4 + 4\sqrt{w} + w$, $4\sqrt{w} = 6 - 2w$, $16w = (6 - 2w)^2$ (squaring), $16w = 36 - 24w + 4w^2$, $4w^2 - 40w + 36 = 0$, $w^2 - 10w + 9 = 0$, $(w - 9)(w - 1) = 0$, or $w = 9, 1$. *Check:* $\sqrt{3w-2} - \sqrt{w} \overset{?}{=} 2$. If $w = 9$, $\sqrt{27-2} - \sqrt{9} = 5 - 3 = 2$. This checks. If $w = 1$, $\sqrt{3-2} - \sqrt{1} = \sqrt{1} - \sqrt{1} = 0$. This does not check. Conclusion: $w = 9$.

71. If $h = 192$ ft, then $192 = 128t - 16t^2$, $16t^2 - 128t + 192 = 0$, $t^2 - 8t + 12 = 0$, or $(t - 6)(t - 2) = 0$; after 2 s and after 6 s. *Note:* The ball goes upward, stops, and turns downward. That is why it reaches 192 ft after 2 s.

72. $A = P(1 + r)^2$, where $A = 1440$ and $P = 1000$. $1440 = 1000(1 + r)^2$, $(1 + r)^2 = 1.44$, $1 + r = \pm\sqrt{1.44}$ (reject the negative!), $1 + r = \sqrt{1.44} = 1.2$, or $r = 0.2$. The rate is 20 percent.

73. Let $x =$ first number. Then $x + 1 =$ second number. If their product is 210, then $x(x + 1) = 210$, $x^2 + x = 210$, $x^2 + x - 210 = 0$, $(x - 14)(x + 15) = 0$, $x = 14, -15$. $x = -15$ is extraneous (it is nonpositive). Thus, $x = 14$, $x + 1 = 15$ (check this!).

74. (A) $4z - 3y \le 0$
(B) $5t \le 3(-y) - 3$ or $5t \le -3y - 3$
(C) $5t > 3 + 3y$
(D) $5t = 3 + 3y$
(E) $5t > 3y$

75. (A) Substituting 3 for x, we get $-x + 5 = -3 + 5 = 2$ and $2 \ge 0$; 3 does satisfy the inequality.
(B) When $x = \dfrac{1}{2}$, $\dfrac{4}{x} + 3 = \dfrac{4}{\frac{1}{2}} + 3 = 11$, and $\dfrac{1}{x} = 2$, so $11 \ge 2$; $\dfrac{1}{2}$ does satisfy the inequality.
(C) When $x = 1$, $x^{-1} + 1 = 1^{-1} + 1 = 2$ and $x^{-2} - 2 = 1^{-2} - 2 = -1$, but $2 > -1$; 1 does not satisfy the inequality.

76. $11t \le -4$, or $t \le -\frac{4}{11}$.

77. $-8s > 2$, or $s < -\frac{2}{8}$, or $s < -\frac{1}{4}$. (Notice the sign reversal!)

78. $-6 < x + 2 < 10$, or $-8 < x < 8$ (adding -2 to all three parts of the inequality).

79. $\left(-\infty, \dfrac{-4}{17}\right]$

80. $(-8, 8)$

81. (A) False; for example, $-5 > -6$, but $25 < 36$.
(B) True. If $a < b$, then when we multiply by b on both sides, we reverse the inequality since $b < 0$.

(C) True; $a > b$ implies $a - b > 0$. Multiplying $a > b$ on both sides by $a - b$ maintains the inequality sign.

(D) True; $a > b$ implies $a - b > 0$, which means $a(a - b) < 0$ and $ab(a - b) > 0$ (note that the signs are reversed).

(E) False; $a - b > 0$, thus $(a - b)^3 > 0$ and $\dfrac{(a-b)^3}{b} < 0$. For example, let $a = -1$ and $b = -2$.

82. (A) $a + b < 12$ ($a, b > 0$). If $a = 1$, then $1 + b < 12$ or $b < 11$, so b can be $1, 2, \ldots , 10$ (10 choices). If $a = 2$, then $2 + b < 12$ or $b < 10$, so b can be $1, 2, \ldots , 9$ (9 choices). Total number of choices = 19.

(B)

$a = 1$	10 choices for b
$a = 2$	9 choices for b
$a = 3$	8 choices for b
$a = 4$	7 choices for b
$a = 5$	6 choices for b
$a = 6$	5 choices for b
$a = 7$	4 choices for b
$a = 8$	3 choices for b
$a = 9$	2 choices for b
$a = 10$	1 choice for b
Total:	55 possibilities

83. Recall that $|x|$ has two common equivalent definitions.

$$|x| = \begin{cases} x & x \geq 0 \\ -x & x < 0 \end{cases} \quad \text{or} \quad |x| = \sqrt{x^2}$$

If $|x| = 6$, then $x = 6$ or $x = -6$. *Check*: $|6| = 6$, $|-6| = 6$. Note that this check works given either of the definitions above: $\sqrt{6^2} = 6$ and $\sqrt{(-6)^2} = 6$.

84. Recall that $|x - a| < b$ can be interpreted as all x whose distance from a is less than b.

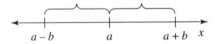

Figure A2.2

In Figure A2.2, notice that the interval $(a - b, a + b)$ satisfies that criterion. In our problem, $a = 0$, $b = 5$. Thus $|x| < 5$ has the solution $-5 < x < 5$ ($-5 = 0 - 5$, $5 = 0 + 5$). More generally, let $a > 0$; then

$\|x\| = a$ has the solution $x = \pm a$
$\|x\| < a$ has the solution $-a < x < a$
$\|x\| > a$ has the solution $x > a$, $x < -a$
$\|cx + d\| < a$ can be written as $-a < cx + d < a$
$\|cx + d\| > a$ can be written as $cx + d > a$, $cx + d < -a$

85. This inequality has no solution: $|x| \geq 0$ for all x.

86. Then $x - 4 = 3$ or $x - 4 = -3$. Then $x = 7$ or $x = 1$.

87. $t - 2 \geq 5$ or $t - 2 \leq -5$. Thus, $t \geq 7$ or $t \leq -3$.

88. $-10 < 2x - 5 < 10$, $-5 < 2x < 15$, or $-\frac{5}{2} < x < \frac{15}{2}$.

89. $-5 < 1 - x < 5$, $-6 < -x < 4$, or $6 > x > -4$. (Careful!)

90. $(x - 2)(x - 1) > 0$. (1) $x - 2 > 0$, $x - 1 > 0$; $x > 2$, $x > 1$, or (2) $x - 2 < 0$, $x - 1 < 0$; $x < 2$, $x < 1$. Simplify: (1) If $x > 2$ and $x > 1$, then $x > 2$, since $x > 2$ guarantees us that $x > 1$. (2) If $x < 2$ and $x < 1$, then $x < 1$, since $x < 1$ guarantees that $x < 2$. Conclusion: $x > 2$ or $x < 1$.

91. $x(3x + 1) > 0$. Then (1) $x > 0$ and $3x + 1 > 0$ or (2) $x < 0$ and $3x + 1 < 0$. Then $x > 0$ and $x > -\frac{1}{3}$, which means $x > 0$, or $x < 0$ and $x < -\frac{1}{3}$, which means $x < -\frac{1}{3}$. Conclusion: $x > 0$ or $x < -\frac{1}{3}$ or, in interval form, $(0, \infty) \cup (-\infty, -\frac{1}{3})$.

92. If $3x^2 < -2x + 1$, then $3x^2 + 2x - 1 < 0$. Thus, $(3x - 1)(x + 1) < 0$ which means (1) $3x - 1 < 0$, $x + 1 > 0$ or (2) $3x - 1 > 0$, $x + 1 < 0$. Then $x < \frac{1}{3}$ and $x > -1$, or $x > \frac{1}{3}$ and $x < -1$ which is impossible. Conclusion: $x < \frac{1}{3}$ and $x > -1$.

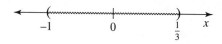

93. Rewrite this equation by isolating one of the radicals, $\sqrt{5x - 1} = 1 + \sqrt{x}$, and square; then $5x - 1 = 1 + 2\sqrt{x} + x$, $4x - 2 = 2\sqrt{x}$, or $2x - 1 = \sqrt{x}$.
Square: $4x^2 - 4x + 1 = x$, $4x^2 - 5x + 1 = 0$, $(4x - 1)(x - 1) = 0$, and $x = \frac{1}{4}$, 1.
Check each prospective root; since we squared the equation, extraneous roots are likely.
When $x = \frac{1}{4}$, $\sqrt{5(\frac{1}{4}) - 1} - \sqrt{\frac{1}{4}} = 0 \neq 1$. $x = 1$ checks and is the only solution.

94. **(A)** The discriminant $b^2 - 4ac = (-8)^2 - 4(1)^2(9) = 28$. Since $28 > 0$ and since 28 is not a perfect square, the roots are irrational and unequal.
  (B) $b^2 - 4ac = (-8)^2 - 4(3)(9) = -44$. Since $-44 < 0$, the roots are imaginary and unequal.
  (C) $b^2 - 4ac = 169$. $169 > 0$ and is a perfect square $(169 = 13^2)$. Thus, the roots are rational and unequal.
  (D) $b^2 - 4ac = (-4\sqrt{3})^2 - 4(4)(3) = 0$. Thus, the roots are real and equal. (*Question:* What are they, and why are they not rational?)

95. Let $x =$ time (hours) required by smaller pipe, $x - 3 =$ time required by larger pipe. Then $\dfrac{1}{x} =$ part filled in 1 h by smaller pipe, $\dfrac{1}{x - 3} =$ part filled in 1 h by larger pipe.

Since the two pipes together fill $\dfrac{\frac{1}{20}}{\frac{3}{3}} = \dfrac{3}{20}$ of the reservoir in 1 h, $\dfrac{1}{x} + \dfrac{1}{x-3} = \dfrac{3}{20}$, $20(x-3) +$

$20x = 3x(x-3)$, $3x^2 - 49x + 60 = (3x-4)(x-15) = 0$, and $x = \frac{4}{3}, 15$. The smaller pipe will fill the reservoir in 15 h and the larger in 12 h.

96. $(x^2 - 6x) + (y^2 - 9y) = -2$, $(x^2 - 6x + 9) + \left(y^2 - 9y + \dfrac{81}{4}\right) = -2 + 9 + \dfrac{81}{4} = \dfrac{109}{4}$,

$(x-3)^2 + \left(y - \dfrac{9}{2}\right)^2 = \dfrac{109}{4}$.

97. (A) $\sqrt{4x^2 - 8x + 9} = 2\sqrt{x^2 - 2x + \dfrac{9}{4}} = 2\sqrt{(x^2 - 2x + 1) + \dfrac{5}{4}} = 2\sqrt{(x-1)^2 + \dfrac{5}{4}}$

(B) $\sqrt{8x - x^2} = \sqrt{16 - (x-4)^2}$

(C) $\sqrt{3 - 4x - 2x^2} = \sqrt{2}\sqrt{\dfrac{3}{2} - 2x - x^2} = \sqrt{2}\sqrt{\dfrac{5}{2} - (x^2 + 2x + 1)} = \sqrt{2}\sqrt{\dfrac{5}{2} - (x+1)^2}$

98. (A) The equation of motion is $s = 120t - 16.1t^2$. When $s = 60$: $60 = 120t - 16.1t^2$ or

$16.1t^2 - 120t + 60 = 0$. $t = \dfrac{120 \pm \sqrt{(120)^2 - 4(16.1)60}}{32.2} = \dfrac{120 \pm \sqrt{10,536}}{32.2} \approx \dfrac{120 \pm 102.65}{32.2} =$

$6.91, 0.54$. After $t = 0.54$ s, the object is 60 ft above the ground and rising. After $t = 6.91$ s, the object is 60 ft above the ground and falling.

(B) The object is at its highest point when $t = \dfrac{-b}{2a} = \dfrac{-(-120)}{2(16.1)} = 3.73$s. Its height is

given by $120t - 16.1 \, t^2 = 120(3.73) - 16.1(3.73)^2 - 223.6$ ft.

99. (A) $b^2 - 4ac = 25 + 24 > 0$; the graph crosses the x axis.

(B) $b^2 - 4ac = 25 - 32 < 0$, and the graph is either wholly above or wholly below the x axis. Since $f(0) > 0$ (the value of the function for any other value of x would do equally well), the graph lies wholly above the x axis.

(C) $b^2 - 4ac = 400 - 400 = 0$; the graph is tangent to the x axis.

(D) $b^2 - 4ac = 4 - 144 < 0$ and $f(0) < 0$; the graph lies wholly below the x axis.

100. (A) $3, \frac{2}{5}$. Here $x_1 + x_2 = \frac{17}{5}$ and $x_1 x_2 = \frac{6}{5}$. The equation is $x^2 - \dfrac{17}{5}x + \dfrac{6}{5} = 0$ or $5x^2 - 17x + 6 = 0$.

(B) $-2 + 3\sqrt{5}, -2 - 3\sqrt{5}$. Here $x_1 + x_2 = -4$ and $x_1 x_2 = 4 - 45 = -41$. The equation is $x^2 + 4x - 41 = 0$.

(C) $\dfrac{3 - i\sqrt{2}}{2}, \dfrac{3 + i\sqrt{2}}{2}$. The sum of the roots is 3, and the product is $\frac{11}{4}$. The equation is $x^2 - 3x + \frac{11}{4} = 0$ or $4x^2 - 12x + 11 = 0$.

101. (A) $x^2 + 4kx + k + 2 = 0$ has one root 0. Let x be 0, then $k + 2 = 0$ and $k = -2$. The equation is $x^2 - 8x = 0$.

(B) $4x^2 - 8kx - 9 = 0$ has one root the negative of the other. Since the sum of the roots is to be 0, $2k = 0$ and $k = 0$. The equation is $4x^2 - 9 = 0$.

(C) $4x^2 - 8kx + 9 = 0$ has roots whose difference is 4. Denote the roots r and $r + 4$. Then $r + (r + 4) = 2r + 4 = 2k$ and $r(r + 4) = \frac{9}{4}$. Solving for $r = k - 2$ in the first and substituting in the second, we have $(k - 2)(k + 2) = \frac{9}{4}$; then $4k^2 - 16 = 9$ and $k = \pm\frac{5}{2}$. The equations are $4x^2 - 20x + 9 = 0$ and $4x^2 - 20x + 9 = 0$.

Chapter 3: Graphs, Relations, and Functions

102. See Figure A3.1

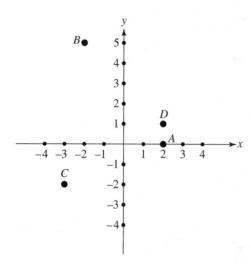

Figure A3.1

103. The distance between two points A and B in the plane $= d(A, B) = AB = \sqrt{(x_2 - x_1)^2 + (y_2 - y_1)^2}$, where A has coordinates (x_1, y_1) and B has coordinates (x_2, y_2). Then, in this case, $d(A, B) = \sqrt{(0-1)^2 + (1-0)^2} = \sqrt{1+1} = \sqrt{2}$.

104. $d(A, B) = \sqrt{[1-(-1)]^2 + (4-5)^2} = \sqrt{2^2 + (-1)^2} = \sqrt{5}$. Notice that since $(x_1 - x_2)^2 = (x_2 - x_1)^2$, the order of subtraction is unimportant.

105. The midpoint M of the segment joining (x_1, y_1) and (x_2, y_2) has coordinates $\left(\dfrac{x_1 + x_2}{2}, \dfrac{y_1 + y_2}{2} \right)$. Thus, in this case M has coordinates $\left(\dfrac{1+3}{2}, \dfrac{2+2}{2} \right) = (2, 2)$. Draw a picture. Does this result make sense?

106. (A)

x axis	y axis	Origin
$-y = x$	$y = -x$	$-y = -x$
Not symmetric	Not symmetric	Symmetric

(B)

x axis	y axis	Origin
$x^2 + (-y)^2 = 4$	$(-x)^2 + y^2 = 4$	$(-x)^2 + (-y)^2 = 4$
$x^2 + y^2 = 4$	$x^2 + y^2 = 4$	$x^2 + y^2 = 4$
Symmetric	Symmetric	Symmetric

(C) x axis	y axis	Origin
$-y = \lvert 2x \rvert$ or $y = -\lvert 2x \rvert$	$y = \lvert 2(-x) \rvert$ or $y = \lvert -2x \rvert$	$-y = \lvert 2(-x) \rvert$ $-y = \lvert 2x \rvert$
Not symmetric	Symmetric	Not symmetric

(D) x axis	y axis	Origin
$-y = \lvert x \rvert + 1$	$y = \lvert -x \rvert + 1$ or $y = \lvert x \rvert + 1$	$-y = \lvert -x \rvert + 1$ $-y = \lvert x \rvert + 1$
Not symmetric	Symmetric	Not symmetric

(E) x axis	y axis	Origin
$-y = x^3$	$y = (-x)^3$ or $y = -x^3$	$-y = (-x)^3$ $y = x^3$
Not symmetric	Not symmetric	Symmetric

107. See Figure A3.2 and question 106 D. Notice that we calculated y only for positive x values since we already knew the graph had y axis symmetry.

x	0	1	2
y	1	2	3

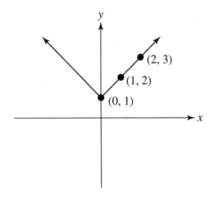

Figure A3.2

108. (A) Figure 3.1 is a function. No two ordered pairs have the same abscissa. Notice (see Figure A3.3) that any line drawn at x_0 perpendicular to the x axis on the x axis intersects the function only once.

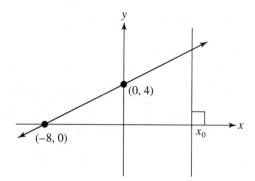

Figure A3.3

(B) Be careful here! Figure 3.2 exhibits a function. For each x value, there is only one y value; thus, no two ordered pairs have the same abscissa. However, unlike question 108A, two ordered pairs share an ordinate, for example, (1, 1) and (−1, 1).

(C) Figure 3.3 is a function. All ordered pairs have different abscissas. Notice that this graph reminds us of $y = x^3$.

(D) Figure 3.4 shows a function. Draw a perpendicular to the x axis at any point on the x axis, and note that it intersects the function only once.

(E) Figure 3.5 is not a function. Since (1, 1) and (1, −1) have the same first element, it violates the definition of a function.

109. (A) f is a function. For every x, $2x − 10$ has only one value. It is impossible to get two values of f from one x.

(B) f is a function. If we plug in a value for x, we get only one value for f.

(C) This equation does define a function. For each x, $|x|$ has only one value.

(D) Remember that x is the independent variable. But for $x = 1$ (for example), if $1 = |y|$, then $y = \pm 1$. Thus, (1, 1) and (1, −1) are both in this relation. The relation is not a function.

(E) This is not a function. Two ordered pairs share an x value, for example, (8, 2) and (8, −2).

110. (A) $F(2) - g(3) = (2^3 - 2 + 4) - (3^2 - 9) = 10 - 0 = 10.$

(B) $F(1) + F(2) = (1^3 - 1 + 4) + (2^3 - 2 + 4) = 4 + 10 = 14.$

(C) $F(0) \cdot f(0) = (0^3 - 0 + 4)(0 - 9) = 4(-9) = -36.$

(D) $\dfrac{f(0)}{F(0)} = \dfrac{0 - 9}{0^3 - 0 + 4} = -\dfrac{9}{4}.$

111. (A) For any real value x, $2x + 5 = y$ is real. Since any real number can replace x, the domain is the set of all real numbers $\mathcal{R}$.

(B) Since any real number squared yields a number, y is defined for any real x. The domain equals the set of all real numbers.

(C) We must ensure that $3 - x^2 \geq 0$. Then $3 \geq x^2$ or $x^2 \leq 3$. Then $-\sqrt{3} \leq x \leq \sqrt{3}$ is the domain.

(D) For any real x, $|x| - 1$ is real. The domain equals the set of all real numbers.

(E) We need to be certain that $2 - x \neq 0$. The domain is the set of all real numbers except for $x = 2$, that is, $(-\infty, 2) \cup (2, \infty)$ or $\mathcal{R} - \{2\}$.

(F) Since x is the independent variable, we need to solve for y to examine the domain. If $x^2 + y^2 = 8$, then $y^2 = 8 - x^2$ or $y = \pm\sqrt{8 - x^2}$. Then $8 - x^2 \geq 0$ if and only if $8 \geq x^2$ or $x^2 \leq 8$; that is, $-2\sqrt{2} < x < 2\sqrt{2}$ is the domain of this relation. Note that this relation is not a function.

(G) The domain is $\mathcal{R}$. For any real x, y is the value 3 and is therefore always defined.

(H) The domain is the set of x values; in this case, the domain = $\{1, 2, 3\}$.

112. (A) The range is the set of ordinates, which in this case is $\{0, 1, a\}$.

(B) Any real y can be expressed as $2x - 3$ for some real x. For example, if $-4 = 2x - 3$, $2x = -1$, or $x = -\frac{1}{2}$, the range = $\mathcal{R}$.

(C) For every $x \in \mathcal{R}$, $x^2 \geq 0$. The range is the nonnegative real numbers, or $\mathcal{R}^+ \cup \{0\}$ or $[0, \infty)$.

(D) $|x| \geq 0$; thus, $-|x| \leq 0$ and $2 - |x| \leq 2$. Range = $(-\infty, 2]$.

(E) $|x| \geq 0$; thus, $|x| + 5 \geq 5$. Range = $[5, \infty)$.

(F) If $x = 2$, $f(x) = 0$. If $x < 2$, $f(x) > 0$. (Note that x cannot be > 2.) The range is $[0, \infty)$.

(G) Since 4 is the function's value for any x, the range is $\{4\}$.

(H) $x \neq 0$. For $x > 0$, $0 < \dfrac{1}{x} < \infty$. For $x < 0$, $-\infty < \dfrac{1}{x} < 0$. The range is $(-\infty, 0) \cup (0, \infty)$.

113. (A) A function is one-to-one if whenever $x_1 \neq x_2$, $f(x_1) \neq f(x_2)$. In this case, if $x_1 \neq x_2$, $3x_1 + 4 \neq 3x_2 + 4$, so the function is one-to-one.

(B) $|x| + 1$ has the same value for x and $-x$. Thus, f is not one-to-one.

(C) $3\sqrt{1 - x}$ is not the same value for two different x choices. This function is one-to-one.

(D) This is one-to-one. No two ordered pairs have the same second element.

(E) This function is not one-to-one. Notice that $(2, 4)$ and $(3, 4)$ have the same second element.

114. (A) Yes; for every x value there is exactly one y value.

(B) Notice the horizontal segments. They occur on $[d, c]$ and $[a, b]$. These are the only intervals over which $y = f(x)$ is constant.

(C) Approaching d from the left, we see that the y values are decreasing. From c to a, they are increasing, and from b on (infinitely far!), f is increasing. Decreasing: $(-\infty, d)$. Increasing: (c, a), (b, ∞).

(D) Nonincreasing means decreasing or constant; nondecreasing means increasing or constant. Nonincreasing: $(-\infty, c]$. Nondecreasing: $[d, \infty)$.

115. See Figure A3.4. This function is of the form $f(x) = ax + b$, which is the form for a linear function. Since we know the graph is a straight line, we find two points. Find a

third point as well to make certain you did not make an error. The third point must lie on the line.

x	0	1
y	-3	-1

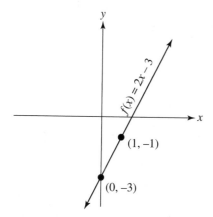

Figure A3.4

116. See Figure A3.5. Notice that since $f(-x) = x^2 = f(x)$, this function's graph is symmetric about the y axis. Also $f(x) \geq 0$ for all x. This graph is a parabola. See Chapter 10 for more about the conic sections in general.

x	-2	-1	0	1	2
y	4	1	0	1	4

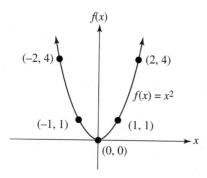

Figure A3.5

117. See Figure A3.6. This function is symmetric about the origin, since $y = x^3$ and $-y = (-x)^3$.

x	0	1	−1	2	−2
y	0	1	−1	8	−8

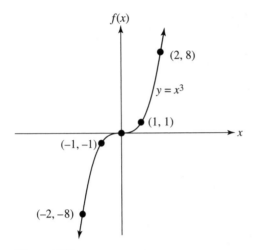

Figure A3.6

118. Figure A3.7 exhibits none of the three symmetries. Notice that $f(x) \geq 0$ for all x and that $x + 2 \geq 0$ or $x \geq -2$. No part of the graph can exist in the shaded areas. Notice that the graph is one-half of a parabola. The "negative branch" is "missing" because of the square root.

x	−2	−1	0	2
$f(x)$	0	1	$\sqrt{2}$	2

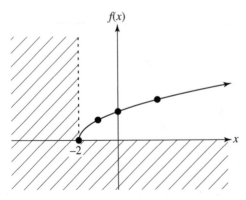

Figure A3.7

119. Figure A3.8 is symmetric about the y axis. Also $f(x) \geq 0$ for all x.

x	-1	-0	1	2	-2
$f(x)$	1	0	1	2	2

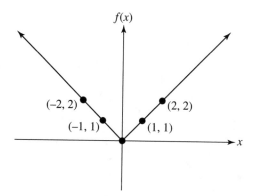

Figure A3.8

120. See Figure A3.9. Here f is the greatest-integer function: For every x, $f(x)$ is the "greatest integer that is not greater than x." Then $[0] = 0$, $[1] = 1$, $[-1] = -1$, $[2] = 2$, $[-2] = -2$, etc. However, $[\frac{1}{2}] = 0$, $[\frac{7}{8}] = 0$, $[4.2] = 4$, $[-2.5] = -3$. The graph then looks as follows:

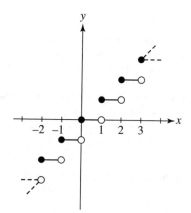

Figure A3.9

121. See Figure A3.10.

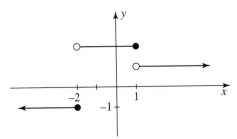

Figure A3.10

122. See Figure A3.11.

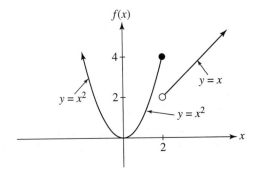

Figure A3.11

123. **(A)** This function is continuous everywhere. There are no breaks in the graph of this function. Until we study continuity more formally, we will use this simple criterion.

(B) This function is continuous everywhere. Clearly there are no breaks in the graph of this function.

(C) In general, all quadratic and linear functions are continuous everywhere. This is a quadratic function (or linear if $a = 0$). It is continuous everywhere.

(D) Without looking at a graph, we note that the place at which we might have a discontinuity is $x = 1$. Since $1 \neq 0$, there is clearly going to be a jump at $x = 1$.

124. Since when $y = mx + b$, $m =$ slope and $b = y$ intercept, in this case m (slope) $= 3$ and b (y intercept) $= 1$.

125. If $-3y = x + 6$, then $y = \dfrac{-x}{3} - 2$. Slope $= -\frac{1}{3}$, y intercept $= -2$.

126. Then $y = mx + b$ where $m = 0$, $b = 3$. Note that this is a horizontal line.

127. This is a vertical line; it has no slope and no y intercept.

128. See Figure A3.12. After plotting the y intercept, make use of the fact that the slope is 3. That means that $\dfrac{(y_2 - y_1)}{(x_2 - x_1)} = 3$ for any (x_1, y_1), (x_2, y_2) on the line. Thus, if x changes by 1, then y changes by 3. In the graph we went from $(0, 2)$ to $(1, 5)$. Therefore, $y_2 - y_1 = 5 - 2$ is a change of 3, and $x_2 - x_1 = 1 - 0$ is a change of 1.

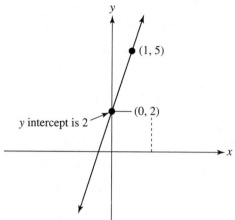

Figure A3.12

129. See Figure A3.13. Notice that $b = 1$ and $m = -1$; then for each change of 1 in x, the y change is -1. Notice that in the graph we went from $(0, 1)$ to $(1, 0)$. Therefore, $y_2 - y_1 = 0 - 1$ is a change of -1, and $x_2 - x_1 = 1 - 0$ is a change of 1.

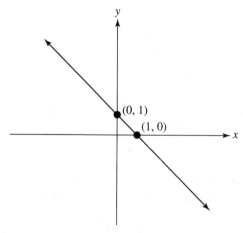

Figure A3.13

130. "No slope" signals a vertical line. This one is $x = 7$.

131. $m = \dfrac{3 - (-1)}{2 - 1} = \dfrac{4}{1} = 4$. Thus, $y - 3 = 4(x - 2)$, $y = 4x - 5$.

132. If the line is parallel to $y = x - 3$, then its slope $m = 1$. Thus, $y - y_1 = m(x - x_1)$, or $y - 1 = 1(x - 2)$, $y = x - 1$.

133. $2x + y = 3$ is the equivalent of a line with slope -2. The slope of our line here must be the negative reciprocal of -2, or $\dfrac{1}{2}$. Thus, $y - 1 = \dfrac{1}{2}(x - 3)$, or $y = \dfrac{x}{2} - \dfrac{1}{2}$.

134. The line must be of the form $x = a$. Since $x = 7$ is on the line, the line is $x = 7$.

135. The line must be of the form $y = a$. Since $y = -6$ is on the line, the line is $y = -6$.

136. Since $y = 2x + 1$ and $y = x - 3$, they will meet when $2x + 1 = x - 3$. Then $x = -4$; thus, since $y = x - 3$, $y = -4 - 3 = -7$. The intersection point is $(-4, 7)$.

137. Since $x^2 = x$ when the graphs intersect, $x^2 - x = 0$, $x(x - 1) = 0$, $x = 0$, $x = 1$. When $x = 0$, $y = 0$. When $x = 1$, $y = 1$. Thus, $(0, 0)$ and $(1,1)$ are the intersection points.

138. $(f + g)(x) = x^2 + (x - 1) = x^2 + x - 1$; $(f - g)(x) = x^2 - (x - 1) = x^2 - x + 1$. Domain of both $= \Re$.

139. $(f + g)(x) = \dfrac{1}{x - 1} + \sqrt{x}$; $(f - g)(x) = \dfrac{1}{x - 1} - \sqrt{x}$. Domain of both $= \{x \in \Re | x \geq 0, x \neq 1\}$.

140. $(f + g)(x) = \sqrt{x + 1} + \sqrt{x - 1}$; $(f - g)(x) = \sqrt{x + 1} - \sqrt{x - 1}$. Domain of both $= \{x \in \Re | x \geq 1\}$. Note that if $x < 1$, $\sqrt{x - 1}$ is not defined.

141. $fg(x) = (x - 3)\sqrt{x + 1}$; $\dfrac{f}{g}(x) = \dfrac{\sqrt{x + 1}}{x - 3}$ $(x \neq 3)$. Domain $fg = \{x \in \Re | x \geq 1\}$. Domain $\dfrac{f}{g} = \{x \in \Re | x \geq -1 \text{ and } x \neq 3\}$.

142. $f \circ g(x) = f(2x + 5) = (2x + 5)^2 = 4x^2 + 10x + 25$.

143. $f \circ g(x) = f\left(\dfrac{1}{2 + x}\right) = \dfrac{1}{1 + \frac{1}{(2+x)}} = \dfrac{2 + x}{3 + x}$.

144. $g \circ f(x) = g(x^2) = 2(x^2) + 5 = 2x^2 + 5$. See question 142.

145. $g \circ f(x) = g\left(\dfrac{1}{1 + x}\right) = \dfrac{1}{2 + \frac{1}{(1+x)}} = \dfrac{1 + x}{3 + 2x}$. See question 143.

146. (A) First note that f is one-to-one, so we are assured of the existence of f^{-1}. Also the domain of f = range of f^{-1}, and the range of f = domain of f^{-1}. Since the domain f = range f = $\Re$, the domain f^{-1} = range f^{-1} = $\Re$.

(B) $y = 2x - 4$. Reverse the x and y: $x = 2y - 4$, $2y = x + 4$, or $y = \dfrac{x}{2} + 2$, $f^{-1}(x) = \dfrac{x}{2} + 2$.

(C) $f^{-1} \circ f(x) = f^{-1}(2x - 4) = \dfrac{2x - 4}{2} + 2 = x - 2 + 2 = x$. Thus, $f^{-1} \circ f(x) = x$.

(D) $f \circ f^{-1}(x) = f\left(\dfrac{x}{2} + 2\right) = 2\left(\dfrac{x}{2} + 2\right) - 4 = x + 4 - 4 = x$. Thus, $f \circ f^{-1}(x) = x$.

(E) See Figure A3.14.

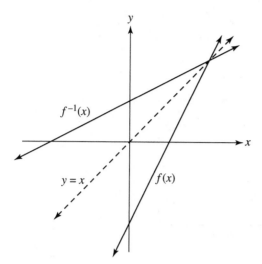

Figure A3.14

147. **(A)** Notice first that $f(x)$ here is one-to-one. Note that if we did not add the restriction $x \geq 0$, f would not be one-to-one and we would not have an inverse function. The domain of f^{-1} = range of f = $[1, \infty)$ since $x^2 + 1 \geq 1$ for all x. The range of f^{-1} = domain of f = $[0, \infty)$ since we are given that $x \geq 0$.

   (B) $y = x^2 + 1; x \geq 0$. Thus if $x = y^2 + 1, y \geq 0$, then $y^2 = x - 1$, or $y = \sqrt{x-1} = f^{-1}(x)$.

   (C) $f^{-1} \circ f(x) = f^{-1}(x^2 + 1) = \sqrt{x^2 + 1 - 1} = \sqrt{x^2} = |x|$. But $x \geq 0$, so $|x| = x$.

   (D) $f \circ f^{-1}(x) = f(\sqrt{x-1}) = (\sqrt{x-1})^2 + 1 = x - 1 + 1 = x$.

   (E) See Figure A3.15.

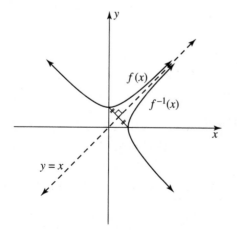

Figure A3.15

148. From Figure A3.16 we see that the base of the box has dimensions $20 - 2x$ by $32 - 2x$ in and the height is x in. Then $V = x\,(20 - 2x)(32 - 2x) = 4x(10 - x)(16 - x)$.

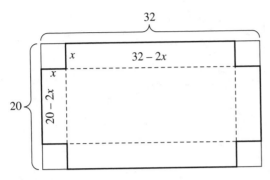

Figure A3.16

149. The dimensions of the field are x and y ft (see Figure A3.17), where $3x + 2y = 600$. Then $y = \frac{1}{2}(600 - 3x)$, and the required area is

$$A = xy = x \cdot \frac{1}{2}(600 - 3x) = \frac{3}{2}x(200 - x)$$

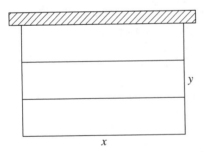

Figure A3.17

150. Let the altitude of the cylinder be denoted by $2h$. From the adjoining Figure A3.18, $h = \sqrt{R^2 - r^2}$ and the required volume is $V = \pi r^2 \cdot 2h = 2\pi r^2 \sqrt{R^2 - r^2}$.

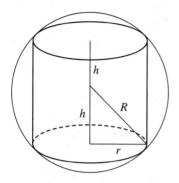

Figure A3.18

151. (A) $f(0, 0) = 2(0)^2 + 3(0)^2 - 4 = -4$.
   (B) $f(2, -3) = 2(2)^2 + 3(-3)^3 - 4 = 31$.
   (C) $f(-x, -y) = 2(-x)^2 + 3(-y)^2 - 4 = 2x^2 + 3y^2 - 4$. Note that $f(x, y) = f(-x, -y)$.

Chapter 4: Polynomial and Rational Functions

152. (A) In standard form, polynomial equations are written from the highest power of x to the lowest. In this case, the standard form is $2x^3 + 4x^2 + 5x - 6 = 0$.
(B) $2x^5 + 0x^4 + x^3 + 0x^2 + 0x + 4 = 0$.
(C) $x^5 + 0x^4 + 0x^3 + 0x^2 + 0x - 2 = 0$.
(D) $(x^2 + 4x + 4) + 5 = 0$, or $x^2 + 4x + 9 = 0$.

153. If $y - k = a(x - h)^2$, then (h, k) is the vertex of the given parabola. In this case, $k = -1$ and $h = 3$. $(3, -1)$ is the vertex.

154. $y = (x^2 - 4x + 4) + 2 = (x - 2)^2 + 2; y - 2 = (x - 2)^2$. The vertex is at $(2, 2)$.

155. (A) If $y = ax^2 + bx + c$, then the parabola opens upward if $a > 0$ and downward if $a < 0$. In this case, $a = 1 > 0$, and the parabola opens upward.
(B) If $y = 2 - x^2$, then $y = -x^2 + 2$ and $a = -1 < 0$; the parabola opens downward.
(C) If $y = x - x^2$, then $y = -x^2 + x$ and $a = -1 < 0$, and the parabola opens downward.
(D) $y = 6 - 5x + x^2$; $a = 1$ and the parabola opens upward.

156. If $y = 2(x - 3)^2 - 1$, then $y + 1 = 2(x - 3)^2$. The vertex is at $(3, -1)$, and the parabola is opening upward (since $a > 0$, where a is the coefficient of x^2). Thus, the function has a minimum value of -1.

157. If $y = -2x^2 + 3$, then $y - 3 = -2(x - 0)^2$. The vertex is at $(0, 3)$, and since $a < 0$, the maximum value is 3.

158. (A) This is a polynomial function, since $y = a_0 x^n + a_1 x^{n-1} + \cdots + a_{n-1} x + a_n$, where n is a positive integer or zero and all a_i are complex numbers.
(B) This is not a polynomial function since $y = x^{-2}$, and -2 is not a positive integer or zero.
(C) This is not a polynomial function. Note that x is the independent variable and thus may not be a positive integer or zero.
(D) This is a polynomial function. If $y = 2x + (4 - 2i)$, then all conditions of the definition in question 158A are satisfied. Note that the constant term $a_n = 4 - 2i$.
(E) This is a polynomial function; the coefficients 2 and $\sqrt{3}$ are complex as in the constant term $- i$.
(F) This is not a polynomial function since. $x^{\frac{1}{2}}$ appears and $\frac{1}{2}$ is nonintegral.

159. The zeros here are 0 with a multiplicity of 2 and 1 with a multiplicity of 1. Thus, the graph touches at 0, and the graph crosses at 1. (Recall that the graph of a polynomial crosses at a zero of odd multiplicity and touches at a zero of even multiplicity.)

160. The zeros here are 1 with a multiplicity of 1 (crosses), -1 with a multiplicity of 1 (crosses), and -2 with a multiplicity of 3 (crosses). (Do you see where 1 and -1 come from? If $x^2 - 1 = 0$, then $x = \pm 1$.)

161. The remainder theorem tells us that the remainder upon division by $x - r$ will be $f(r)$. In this case, $r = -2$, so $f(-2) = (-2)^3 - 5(-2)^2 - 3(-2) + 15 = -7$.

162. By the factor theorem, we must show that r is a root of $f(x) = 0$ to show that $x - r$ is a factor of $f(x)$. In this case, $f(2) = 2 \cdot 2^3 - 3 \cdot 2^2 + 2 \cdot 2 - 8 = 0$. Thus, $x - 2$ is a factor.

163. By the factor theorem, $x - 1$, $x - 2$, $x + 3$ must be factors. Then $f(x) = (x - 1)(x - 2)(x + 3) = (x^2 - 3x + 2)(x + 3) = x^3 - 7x + 6 = 0$.

164. See Figure A4.1. If $f(x) = x^3 = 0$, then $x = 0$. The graph crosses the x axis at $x = 0$ since multiplicity is odd. When $x = 0$, $y = 0$, so the graph crosses the y axis when $x = 0$. Since $f(-x) = -f(x)$, the graph is symmetric about the origin. As x grows larger, so does y; as x grows smaller, so does y. Notice the steps we have gone through here. We will use them throughout this section.

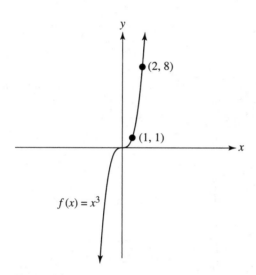

Figure A4.1

165. See Figure A4.2. If $(x - 3)(x^2 - 1) = 0$, then $x = 3, \pm 1$. Each of these zeros has a multiplicity of 1, so the graph will cross the x axis at these points. Also the three symmetry tests fail. Investigating the behavior of this function, we find that (*a*) it is negative for $x < -1$, (*b*) it is positive for $-1 < x < 1$, (*c*) it is negative for $1 < x < 3$, and (*d*) it is positive for $x > 3$.

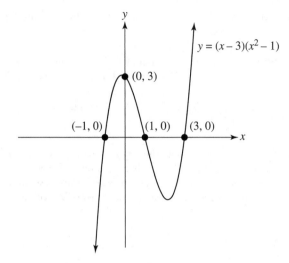

Figure A4.2

166. See Figure A4.3. This graph has four zeros: $x = \pm 1$, $x = \pm 3$, all with odd multiplicity. Also replacing x with $-x$ does not change the equation, so the graph exhibits y-axis symmetry. $f(0) = 9$. $f > 0$ for $x < -3$, $x > 3$, $-1 < x < 1$; and $f < 0$ for $-3 < x < -1$, $1 < x < 3$.

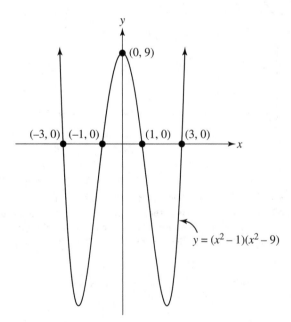

Figure A4.3

167. **(A)** Volume $= l \cdot w \cdot h = x \cdot x \cdot (108 - x - x - x - x) = x^2(108 - 4x)$.

(B) $108 - 4x$ must be nonnegative in order that $V \geq 0$. Thus, $0 \leq x \leq 27$. If $x = 0$ or $x = 27$, the volume of the box is 0.

(C) See Figure A4.4.

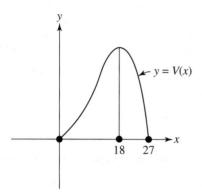

Figure A4.4

(D) To the nearest inch, the maximum occurs when $x \approx 18$. $V(18) = 18 \cdot 18 \cdot 36 = 11{,}664 \text{ in}^3 \approx$ maximum volume.

168. **(A)** Figure 4.6. Note that if $a > b$, then $f(a) > f(b)$ for all a, b. None of the other graphs have this property. The graph of Figure 4.6 is a polynomial of degree 1.

(B) Figures 4.1 and 4.2. For both of these we see that if (x, y) is on the graph, so is $(-x, y)$.

(C) Only Figure 4.4. Notice that if (x, y) is on the graph, so is $(x, -y)$.

(D) Figure 4.5. Notice that the graph crosses the x axis four times. Each zero must have odd multiplicity for this to happen.

(E) In Figure 4.3, $f > 0$ when x is in the intervals $(-1, \frac{1}{2})$ and $(1, \infty)$. In Figure 4.5, $f > 0$ when x is in the intervals $(-2, -1)$ and $(0,3)$.

169.

$$
\begin{array}{cccc|c}
1 & 1 & 1 & 3 & \underline{1} \\
& 1 & 2 & 3 & \\
\hline
\underbrace{1 + 2 + 3}_{\text{Quotient}} & & \underbrace{+6}_{\text{Remainder}} &
\end{array}
$$

The quotient is $1x^2 + 2x + 3 = x^2 + 2x + 3$, and the remainder is 6.

Outline of method:

(A) Write down the coefficients of the dividend from left to right in decreasing order of the powers of x. Insert 0 for any missing terms.

$$1 \quad 1 \quad 1 \quad 3$$

(B) To the right of this, write the a of the divisor $x - a$.

$$1 \quad 1 \quad 1 \quad 3 \quad \underline{1}$$

(C) On the *third* line (skip one line) rewrite the leading coefficient.

$$1 \quad 1 \quad 1 \quad 3 \ \underline{\lfloor 1}$$

$$\overline{\hspace{3cm}}$$

$$1$$

(D) Multiply this leading coefficient by a, and write it on the second line in the second column as shown. Then add the terms in column 2, and write the sum on the third line as shown.

$$1 \quad 1 \quad 1 \quad 3 \ \underline{\lfloor 1}$$

$$1$$

$$\overline{\hspace{3cm}}$$

$$1 \quad 2$$

(E) Multiply the sum obtained in (D) by a, and repeat the procedure in (D) for each column.

$$1 \ 1 \ 1 \qquad 3 \ \underline{\lfloor 1}$$

$$1 \ 2 \qquad 3$$

$$\overline{\hspace{4cm}}$$

$$\underbrace{1 \ 2 \ 3}_{\text{Coefficients of quotient}} \quad \underbrace{6}_{\text{Remainder}}$$

The quotient is $(1)x^2 + (2)x + 3 = x^2 + 2x + 3$, and the remainder is 6.

170. We will divide by $x + 2$; the remainder will be $f(-2)$.

$$
\begin{array}{rrrr|r}
1 & -7 & 12 & -3 & \underline{-2} \\
  & -2 & 18 & -60 & \\
\hline
1 & -9 & 30 & -63 &
\end{array}
$$

The remainder is $-63 = f(-2)$.

171. Divide by $x - 3$.

$$
\begin{array}{rrrr|r}
1 & -7 & 12 & -3 & \underline{3} \\
  & 3 & -12 & 0 & \\
\hline
1 & -4 & 0 & -3 &
\end{array}
$$

The remainder is $-3 = f(3)$.

172. We will divide and show that we obtain a zero remainder.

$$
\begin{array}{rrrr|r}
1 & 0 & -18 & 27 & \underline{3} \\
  & 3 & 9 & -27 & \\
\hline
1 & 3 & -9 & 0 &
\end{array}
$$

The remainder is 0, so $x - 3$ is a factor.

173.

$$
\begin{array}{r}
\begin{array}{rrrrr}
1 & 0 & -\frac{3}{4} & \frac{1}{2} & \frac{3}{8} \\
 & -\frac{1}{2} & \frac{1}{4} & \frac{1}{4} & -\frac{3}{8}
\end{array} \quad \lfloor -\frac{1}{2} \\
\hline
\begin{array}{rrrrr}
1 & -\frac{1}{2} & -\frac{1}{2} & \frac{3}{4} & 0
\end{array}
\end{array}
$$

The remainder is 0, so $x + \frac{1}{2}$ is a factor.

174. Every polynomial $P(x)$ of degree $d \geq 1$ with complex coefficients has at least one complex zero. (Remember! Real numbers are also complex.)

175. $x(x-1)(x-2) = x(x^2 - 3x + 2) = 0$, or $x^3 - 3x^2 + 2x = 0$.

176. Then $x - (1 + \sqrt{3})$, $x - (1 - \sqrt{3})$, $x - (-1 - \sqrt{3})$ are the three factors.

$[x - (1 + \sqrt{3})][x - (1 - \sqrt{3})][x - (-1 - \sqrt{3})] = x^3 - (1 - \sqrt{3})\, x^2 - (4 + 2\sqrt{3})\, x - (2\sqrt{3} + 2)$ in polynomial form.

177. $P(x) = (x + 4)^2 [x - (2 - 3i)][x - (2 + 3i)]$ where the degree of $P(x) = 4$.

178. If $(x - a)^m$ is a factor of $P(x)$, then a is a zero. In this case, $x - 0$, $x - 2$, $x + 3$ are factors, so $0, 2, -3$ are zeros.

179. $x = -i$, 1, and -4 are all zeros.

180. If $2x - 1 = 0$, $P(x) = 0$. If $2x + 5 = 0$, $P(x) = 0$. Thus, the zeros are $x = \frac{1}{2}, -\frac{5}{2}, 3$.

181. If 1 is a zero, then $x - 1$ is a factor. Using synthetic division.

$$
\begin{array}{r}
\begin{array}{rrrr}
1 & -3 & 1 & 1 \\
 & 1 & -2 & -1
\end{array} \quad \lfloor 1 \\
\hline
\begin{array}{rrrr}
1 & -2 & -1 & 0
\end{array}
\end{array}
$$

$x^2 - 2x - 1$ is a factor; but if $x^2 - 2x - 1 = 0$, then $x = \dfrac{-b \pm \sqrt{b^2 - 4ac}}{2a} = 1 \pm \sqrt{2}$. Thus, the remaining zeros are $1 + \sqrt{2}$ and $1 - \sqrt{2}$.

182. Then $1 - \sqrt{3}$ must be a zero, so $x - (1 - \sqrt{3})$ is a factor of $P(x)$ as well.

$[x - (1 + \sqrt{3})][x - (1 - \sqrt{3})] = x^2 - 2x - 2$. Since $(x^3 - 3x^2 + 2) \div (x^2 - 2x - 2) = x - 1$, 1 and $1 - \sqrt{3}$ are the other zeros.

183. If $-1 - 3i$ is a zero, then $-1 + 3i$ is as well. $[x - (-1 + 3i)][x - (-1 - 3i)] = x^2 + 2x + (1 - 9i^2) = x^2 + 2x + 10$. We divide $x^4 - x^3 + 6x^2 - 26x + 20$ by $x^2 - 2x + 10$ and obtain $x^2 - 3x + 2 = (x - 2)(x - 1)$. Thus, the other zeros are $2, 1, -1 + 3i$.

184. If $\frac{p}{q}$ (a rational number in lowest terms) is a zero of $P(x)$, then $p \mid a_0$ and $q \mid a_n$, where a_0 is the constant term. In this case, any zero $\frac{p}{q}$ must be such that $p \mid 2$ (p divides the constant term) and $q \mid 2$ (q divides the coefficient of x^2). Thus, $p = \pm 1, \pm 2$; $q = \pm 1, \pm 2$; and $\frac{p}{q} = \pm 1, \pm 2, \pm \frac{1}{2}$.

185. The possible zeros are $\pm 2, \pm 1$.

$$
\begin{array}{rrrr|r}
1 & 3 & 1 & -2 & \underline{-2} \\
  & -2 & -2 & 2 & \\
\hline
1 & 1 & -1 & 0 & \\
\end{array}
$$

Thus, -2 is a zero. Checking with synthetic division, we find that 2 and ± 1 do *not* yield a zero remainder. Thus, -2 is the only rational zero.

186. Possible rational zeros are $\pm \frac{1}{2}, \pm 2, \pm 1$. Using synthetic division, we find $-\frac{1}{2}, -1, 2$ to be rational zeros.

$$
\begin{array}{rrrr|r}
2 & -1 & -5 & -2 & \underline{-\frac{1}{2}} \\
  & -1 & 1 & 2 & \\
\hline
2 & -2 & -4 & 0 & \\
\end{array}
$$

Check the other roots, using synthetic division.

187. The possible rational roots are $\pm 1, \pm 2$. Using synthetic division, we find $-2, 1$ to be roots, since $(x^4 - x^3 - 5x^2 + 3x + 2) \div (x + 2) = x^3 - 3x^2 + x + 1$, and $(x^3 - 3x^2 + x + 1) \div (x - 1) = x^2 - 2x - 1$.

$$
\begin{array}{rrrrr|r}
1 & -1 & -5 & 3 & 2 & \underline{-2} \\
  & -2 & 6 & -2 & -2 & \\
\hline
1 & -3 & 1 & 1 & 0 & \\
\end{array}
\qquad
\begin{array}{rrrr|r}
1 & -3 & 1 & 1 & \underline{1} \\
  & 1 & -2 & -1 & \\
\hline
1 & -2 & -1 & 0 & \\
\end{array}
$$

If $x^2 - 2x - 1 = 0$, then $x = 1 \pm \sqrt{2}$. The roots are $1 \pm \sqrt{2}, -2, 1$.

188. $P(x) = 2x^2 + x - 4$ which has *one* variation of sign (+ to −). $P(-x) = 2x^2 - x - 4$ which has *one* variation of sign (+ to −1). Descartes' rule of signs tells us that there will be 1 positive zero [one variation in sign in $P(x)$] and 1 negative zero [one variation in sign in $P(-x)$]. Note that the other possibility by Descartes' rule does not apply here. It is not possible to subtract an even integer from 1 to get a number of zeros.

189. $S(x)$ has two variations. $S(-x) = -x^5 + x^4 - x^3 - x^2 + 1$ which has three variations. Thus, $S(x)$ has 2 or 0 positive and 3 or 1 negative zeros.

190. $P(3) = 3^2 - 3(3) - 2 = -2$, and $P(4) = 4^2 - 3(4) - 2 = 2$; since $P(3)$ and $P(4)$ are opposite in sign, there must be at least one real zero between them.

191. (A) This is rational, since $f(x) = cx + d = \dfrac{cx + d}{1}$, where $cx + d$ and 1 are polynomials.

(B) This is nonrational. The numerator cannot be written as a polynomial. If we divide, we find $f(x) = \sqrt{x - 2}$.

(C) This is nonrational. $|x|$ is not a polynomial. Look at Figure A4.5 which is the graph of $g(x)$. Do you notice the sharp peak? You will learn in calculus that this ensures that the function is nonrational.

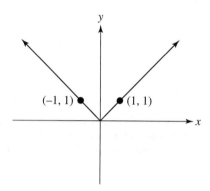

Figure A4.5

(D) 1 and πx are both polynomials. This is a rational function. Do not be misled by the irrational coefficient of x.

(E) This is nonrational. The power to which 3 is raised is variable.

(F) The numerator and denominator are each a polynomial of degree 2. This is a rational function.

(G) This is nonrational. The exponent x is variable.

192. Vertical: Since $f(x) = \dfrac{x}{x+1}$, as $x + 1 \rightarrow 0$, f grows without bound. Thus, $x = -1$ is a vertical asymptote. Horizontal: As x grows without bound, $f(x) = \dfrac{x}{x+1}$ gets nearer and nearer to 1. $y = 1$ is a horizontal asymptote.

193. As $x^2 - 4 \rightarrow 0$, $\dfrac{x^2}{x^2 - 4} \rightarrow \infty$, which implies that $x^2 - 4 = 0$ will give us vertical asymptotes. $x = 2$, $x = -2$ are vertical asymptotes. As $x \rightarrow \infty$, $\dfrac{x^2}{x^2 - 4} \rightarrow 1$. $y = 1$ is a horizontal asymptote.

194. If $x \rightarrow 2$, $x \rightarrow -1$, or $x \rightarrow -2$, the denominator approaches 0. Thus, $x = 2$, $x = -1$, $x = -2$ are vertical asymptotes. As $x \rightarrow \infty$, $f(x) \rightarrow 0$, and $y = 0$ is a horizontal asymptote.

195. As $x \to -1$, $x + 1 \to 0$. $x = -1$ is a vertical asymptote. As $x \to \infty$, $f \to 1$, and $y = 1$ is a horizontal asymptote. We plot the points $(1, \frac{1}{2})$, $(-2, 2)$, $(2, \frac{2}{3})$. See Figure A4.6.

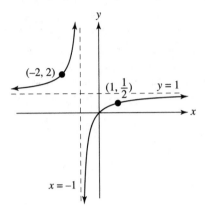

Figure A4.6

196. If $x^2 - 4 = 0$, $x = \pm 2$. As $x \to 2$ or $x \to -2$, f grows without bound; $x = \pm 2$ are vertical asymptotes. As $x \to \infty$, $f \to 1$, so $y = 1$ is a horizontal asymptote. Replacing x by $-x$ leaves the equation line unchanged: the graph exhibits y-axis symmetry. We plot the points $(0, 0)$, $(1, -\frac{1}{3})$, $(3, \frac{9}{5})$. See Figure A4.7.

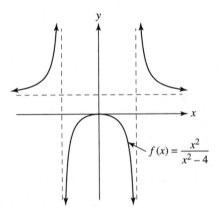

Figure A4.7

Chapter 5: Systems of Equations and Inequalities

197. See Figure A5.1. Graph the two lines $l_1: x + 2y = 5$ and $l_2: 3x - y = 1$. For l_1, if $x = 0$, $y = 2.5$; if $y = 0$, $x = 5$. For l_2, if $x = 0$, $y = -1$; if $y = 0$, $x = \frac{1}{3}$. Next find where these lines intersect. The intersection point is $(1, 2)$. Thus, $x = 1, y = 2$ is the solution. *Check*: $1 + 2(2) = 5$, $3(1) - 2 = 1$.

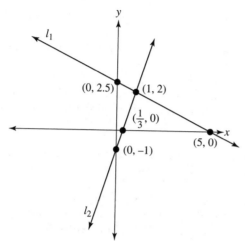

Figure A5.1

198. (A) Recall that if $ax + by = c$ and $dx + ey = f$, then the system is independent if $\dfrac{d}{a} \neq \dfrac{e}{b}$, dependent if $\dfrac{d}{a} = \dfrac{e}{b} = \dfrac{f}{c}$, and inconsistent if $\dfrac{d}{a} = \dfrac{e}{b} \neq \dfrac{f}{c}$. Here $a = 3$, $b = 1$, $c = 6$, $d = 2$, $e = 1$, $f = 5$. Then $\dfrac{d}{a} = \dfrac{2}{3}$; $\dfrac{e}{b} = \dfrac{1}{1} = 1$; thus, $\dfrac{d}{a} \neq \dfrac{e}{b}$, and the system is independent.

(B) $\dfrac{d}{a} = \dfrac{3}{1}$; $\dfrac{e}{b} = \dfrac{-1}{1}$. The system is independent.

(C) $\dfrac{d}{a} = \dfrac{2}{-4}$; $\dfrac{e}{b} = \dfrac{1}{-2}$; $\dfrac{f}{c} = \dfrac{10}{-20}$; $\dfrac{d}{a} = \dfrac{e}{b} = \dfrac{f}{c}$, and the system is dependent.

(D) $\dfrac{d}{a} = \dfrac{2}{-4}$; $\dfrac{e}{b} = \dfrac{-1}{-1}$. The system is independent.

(E) $\dfrac{d}{a} = \dfrac{1}{1}$; $\dfrac{e}{b} = \dfrac{1}{1}$; $\dfrac{f}{c} = \dfrac{5}{8}$. Then $\dfrac{d}{a} = \dfrac{e}{b} \neq \dfrac{f}{c}$, and the system is inconsistent.

(F) Then $x - 3y = 8$ and $-x + 2y = 4$. $\dfrac{d}{a} = \dfrac{1}{-1}$; $\dfrac{e}{b} = -\dfrac{3}{2}$. This system is independent.

(G) $-x + y = 0$ and $-x + 2y = 6$. $\dfrac{d}{a} = \dfrac{-1}{-1}$; $\dfrac{e}{b} = \dfrac{1}{2}$. The system is independent.

199. Multiply the first equation by 3 and add:

$$12x - 3y = 30$$
$$\underline{x + 3y = 9}$$
$$13x \qquad = 39$$

Then $x = 3$. Substituting into $x + 3y = 9$ yields $3 + 3y = 9$, so $3y = 9 - 3 = 6$. Thus, $y = 2$.

200. Multiply the first equation by 2 and subtract:

$$4x + 2y = 10$$
$$\underline{4x + 2y = 10}$$
$$0 = 0$$

The system is dependent.

201. Subtracting, we get $(3x - 3x) + (y - y) = 20 - 30$, or $0 = -10$. The system is inconsistent.

202. If $x + y = 16$, then $x = 16 - y$. Substitute $16 - y$ for x in the second equation. Then $(16 - y) - y = 10$. $16 - 2y = 10$, $2y = 6$, or $y = 3$. Substituting for y, we get $x + 3 = 16$, or $x = 16 - 3 = 13$. Check this.

203. Let $x =$ one number and $y =$ the other. Then $x + y = 20$ and $x - y = 10$. Thus, $2x = 30$, or $x = 15$, and $y = x - 10 = 5$. *Check*: Sum $= 20$; difference $= 10$.

204. Let $l =$ length and $w =$ width. Then $l = 10 + w$ and $2l + 2w = 60$. Substituting for l gives $2(10 + w) + 2w = 60$, $20 + 2w + 2w = 60$, $4w = 40$, or $w = 10$ in. Then $l = 10$ in $+ 10$ in $= 20$ in. *Check*: $20 = 10 + 10$ (length is 10 more), and $P = 2\,(20) + 2\,(10) = 60$.

205. Let $\frac{x}{y}$ be the original fraction. Then

(1) $$\frac{x+2}{y} = \tfrac{1}{4} \quad \text{or} \quad 4x - y = -8$$

(2) $$\frac{x}{y-6} = \tfrac{1}{6} \quad \text{or} \quad 6x - y = -6$$

Subtract (1) from (2): $2x = 2$ and $x = 1$
Substitute $x = 1$ in (1): $4 - y = -8$ and $y = 12$

The fraction is $\frac{1}{12}$.

206. Let $x =$ rate in still water $\left(\dfrac{\text{mi}}{\text{h}}\right)$, $y =$ rate of the river $\left(\dfrac{\text{mi}}{\text{h}}\right)$. Then $x + y =$ rate downstream and $x - y =$ rate upstream.

Now $x + y = 6$ (1)   Add (1) and (2): $2x = 9$ and $x = 4\tfrac{1}{2}$
    $x - y = \tfrac{6}{2} = 3$ (2) Subtract (2) from (1): $2y = 3$ and $y = 1\tfrac{1}{2}$

The rate in still water is $4\tfrac{1}{2}\dfrac{\text{mi}}{\text{h}}$, and the rate of the river is $1\tfrac{1}{2}\dfrac{\text{mi}}{\text{h}}$.

207. Eliminate z:

Rewrite (1): $x - 5y + 3z = 9$ Rewrite (2): $2x - y + 4z = 6$
Multiply (3) by -3: $-9x + 6y - 3z = -6$ Multiply (3) by -4: $-12x + 8y - 4z = -8$
Add: $-8x + y = 3$ (4)   Add: $-10x + 7y = -2$ (5)

Multiply (4) by -7: $56x - 7y = -21$

Rewrite (5): $-10x + 7y = -2$

Add: $46x = -23$

$x = -\tfrac{1}{2}$

Substitute $x = -\tfrac{1}{2}$ in (4):

$-8(-\tfrac{1}{2}) + y = 3$ and $y = -1$

Substitute $x = -\tfrac{1}{2},\ y = -1$ in (1):

$-\tfrac{1}{2} - 5(-1) + 3z = 9$ and $z = \tfrac{3}{2}$.

Check: Using (2), $2(-\tfrac{1}{2}) - (-1) + 4(\tfrac{3}{2}) = -1 + 1 + 6 = 6$.

208. Corresponding elements must be equal. Thus, $a = 1$, $b = 5$, $c = -2$, $d = 3$.

209. A matrix is $m \times n$ if it has m rows (horizontal) and n columns (vertical). In this case, B has two rows and two columns, so it is 2×2; E has one row and four columns, so it is 1×4.

210. $[-2, 0]$ is the second row. Then -2 is in the second-row, first-column position.

211. $A + B = \begin{bmatrix} 1 & -1 \\ 3 & 1 \end{bmatrix} + \begin{bmatrix} -3 & 2 \\ -2 & -3 \end{bmatrix} = \begin{bmatrix} -2 & 1 \\ 1 & -2 \end{bmatrix}$

212. For any matrix B, $-B$ is the matrix such that $B + (-B) =$ zero matrix. All the entries of $-B$ must be the negatives of the corresponding entries in B.

$$\begin{bmatrix} -(-3) & -2 \\ -(-2) & -(-3) \end{bmatrix} = \begin{bmatrix} 3 & -2 \\ 2 & 3 \end{bmatrix}$$

Note that $B + (-B) = 0$ ($0 =$ the zero matrix).

213. $3D - 4C = 3D + (-4C) = \begin{bmatrix} 6 \\ 9 \\ 15 \end{bmatrix} + \begin{bmatrix} 8 \\ 12 \\ -4 \end{bmatrix} = \begin{bmatrix} 14 \\ 21 \\ 11 \end{bmatrix}.$

214. $2B = 2\begin{bmatrix} -5 & 2 & 4 \\ 3 & 0 & -1 \end{bmatrix} = \begin{bmatrix} -10 & 4 & 8 \\ 6 & 0 & -2 \end{bmatrix}$

215. $A^T =$ the transpose of $A =$ the matrix which has rows and columns of A interchanged.

$A^T = \begin{bmatrix} 2 & 3 & 1 \\ 0 & -4 & 5 \end{bmatrix}^T = \begin{bmatrix} 2 & 0 \\ 3 & -4 \\ 1 & 5 \end{bmatrix}$

Row 1 of A is column 1 of A^T.

216. This product is defined since the number of columns of $\begin{bmatrix} 1 & 2 \\ 3 & 1 \end{bmatrix}$ equals the number of rows of $\begin{bmatrix} -1 & 2 \\ -1 & 2 \end{bmatrix}$. We then find the product as follows:

$$\begin{bmatrix} 1 & 2 \\ 3 & 1 \end{bmatrix}\begin{bmatrix} -1 & 2 \\ -1 & 2 \end{bmatrix} = \begin{bmatrix} 1(-1)+2(-1) & 1\cdot 2+2\cdot 2 \\ 3(-1)+1(-1) & 3\cdot 2+1\cdot 2 \end{bmatrix}$$

In general, $(AB)_{m \times n}$ of the matrices $A_{m \times s}$ and $B_{s \times n}$ is the matrix whose ijth element is $a_{i1}b_{1j}+a_{i2}b_{2j}+\cdots+a_{is}b_{sj}$, where i ranges from 1 to m and j ranges from 1 to n. In this case, then, the product is $\begin{bmatrix} -1-2 & 2+4 \\ -3-1 & 6+2 \end{bmatrix} = \begin{bmatrix} -3 & 6 \\ -4 & 8 \end{bmatrix}$.

217. Let $A = \begin{bmatrix} 1 \\ 1 \\ 0 \end{bmatrix}$ and $B = \begin{bmatrix} 2 & 3 & 1 \end{bmatrix}$. Is AB defined? Yes; the number of columns of A = the number of row of B = 1. Then $\begin{bmatrix} 1 \\ 1 \\ 0 \end{bmatrix}\begin{bmatrix} 2 & 3 & 1 \end{bmatrix} = \begin{bmatrix} 1\cdot 2 & 1\cdot 3 & 1\cdot 1 \\ 1\cdot 2 & 1\cdot 3 & 1\cdot 1 \\ 0\cdot 2 & 0\cdot 3 & 0\cdot 1 \end{bmatrix} = \begin{bmatrix} 2 & 3 & 1 \\ 2 & 3 & 1 \\ 0 & 0 & 0 \end{bmatrix}$.

218. The product is $[2 \cdot 2 + 1 \cdot 0 + 3 \cdot 1 \quad 2 \cdot 1 + 1 \cdot 0 + 3 \cdot 0 \quad 2 \cdot 0 + 1 \cdot 0 + 3 \cdot 2] =$ $[7 \quad 2 \quad 6]$.

219. The product is not defined. Do you see why? The number of columns of $\begin{bmatrix} 4 \\ 1 \\ 6 \end{bmatrix} \neq$ the number of rows of $\begin{bmatrix} 1 & 0 & 1 \\ 0 & 2 & 0 \\ 0 & 0 & 1 \end{bmatrix}$. Before going on, compose another example in which the product of two matrices is not defined.

220. The product is $\begin{bmatrix} 1\cdot 3+2\cdot 4+3\cdot 6 & 1\cdot 1+2\cdot 2+3\cdot 0 \\ 1\cdot 3+0\cdot 4+1\cdot 6 & 1\cdot 1+0\cdot 2+1\cdot 0 \end{bmatrix} = \begin{bmatrix} 29 & 5 \\ 9 & 1 \end{bmatrix}$.

221. $\begin{bmatrix} 1 & 6 \end{bmatrix}\cdot\begin{bmatrix} 0 \\ 1 \end{bmatrix} = 1\cdot 0+6\cdot 1 = 6; \begin{bmatrix} 2 & 3 & 4 \end{bmatrix}\cdot\begin{bmatrix} 1 \\ 2 \\ 3 \end{bmatrix} = 2\cdot 1+3\cdot 2+4\cdot 3 = 2+6+12 = 20$. So

$\begin{bmatrix} 1 & 6 \end{bmatrix}\cdot\begin{bmatrix} 0 \\ 1 \end{bmatrix} = \begin{bmatrix} 2 & 3 & 4 \end{bmatrix}\cdot\begin{bmatrix} 1 \\ 2 \\ 3 \end{bmatrix} = 6+20 = 26$.

222. $AB = \begin{bmatrix} 1 \cdot 1 + 2 \cdot 2 & 1 \cdot 1 + 2 \cdot 3 \\ 0 \cdot 1 + 1 \cdot 2 & 0 \cdot 2 + 1 \cdot 3 \end{bmatrix} = \begin{bmatrix} 5 & 7 \\ 2 & 3 \end{bmatrix}$

$BA = \begin{bmatrix} 1 \cdot 1 + 1 \cdot 0 & 1 \cdot 2 + 1 \cdot 1 \\ 2 \cdot 1 + 3 \cdot 0 & 2 \cdot 2 + 3 \cdot 1 \end{bmatrix} \neq \begin{bmatrix} 5 & 7 \\ 2 & 3 \end{bmatrix}$

$AB \neq BA$

223. $AB = \begin{bmatrix} 5 & 7 \\ 2 & 3 \end{bmatrix}$ (See question 222)

$(AB)C = \begin{bmatrix} 5 & 7 \\ 2 & 3 \end{bmatrix}\begin{bmatrix} -3 & 1 \\ -1 & 2 \end{bmatrix} = \begin{bmatrix} 5(-3) + 7(-1) & 5 \cdot 1 + 7 \cdot 2 \\ 2(-3) + 3(-1) & 2 \cdot 1 + 3 \cdot 2 \end{bmatrix} = \begin{bmatrix} -22 & 19 \\ -9 & 8 \end{bmatrix}$

$BC = \begin{bmatrix} 1 & 1 \\ 2 & 3 \end{bmatrix}\begin{bmatrix} -3 & 1 \\ -1 & 2 \end{bmatrix} = \begin{bmatrix} -4 & 3 \\ -9 & 8 \end{bmatrix}$

$A(BC) = \begin{bmatrix} 1 & 2 \\ 0 & 1 \end{bmatrix}\begin{bmatrix} -4 & 3 \\ -9 & 8 \end{bmatrix} = \begin{bmatrix} -22 & 19 \\ -9 & 8 \end{bmatrix} = (AB)C$

224. $A(B+C) = \begin{bmatrix} 1 & 2 \\ 0 & 1 \end{bmatrix}\begin{bmatrix} -2 & 2 \\ 1 & 5 \end{bmatrix} = \begin{bmatrix} 0 & 12 \\ 1 & 5 \end{bmatrix}$. $AB = \begin{bmatrix} 5 & 7 \\ 2 & 3 \end{bmatrix}$. $AC = \begin{bmatrix} -5 & 5 \\ -1 & 2 \end{bmatrix}$.

$AB + AC = \begin{bmatrix} 0 & 12 \\ 1 & 5 \end{bmatrix} = A(B+C)$.

225. $B + C = \begin{bmatrix} -2 & 2 \\ 1 & 5 \end{bmatrix}$. $(B+C)A = \begin{bmatrix} -2 & 2 \\ 1 & 5 \end{bmatrix}\begin{bmatrix} 1 & 2 \\ 0 & 1 \end{bmatrix} = \begin{bmatrix} -2 & -2 \\ 1 & 7 \end{bmatrix}$. $BA = \begin{bmatrix} 1 & 3 \\ 2 & 7 \end{bmatrix}$ See

question 222. $CA = \begin{bmatrix} -3 & 1 \\ -1 & 2 \end{bmatrix}\begin{bmatrix} 1 & 2 \\ 0 & 1 \end{bmatrix} = \begin{bmatrix} -3 & -5 \\ -1 & 0 \end{bmatrix}$. $BA + CA = \begin{bmatrix} -2 & -2 \\ 1 & 7 \end{bmatrix} = (B+C)A$.

226. Remember that $\begin{vmatrix} a_{11} & a_{12} \\ a_{21} & a_{22} \end{vmatrix} = a_{11}a_{22} - a_{21}a_{12}$. In this case $\begin{vmatrix} 1 & 6 \\ 5 & 4 \end{vmatrix} = 1 \cdot 4 - 5 \cdot 6 = 4 - 30 = -26$.

227. Using row 1, we get

$$1(-1)^{1+1}\begin{vmatrix} 1 & -2 \\ 1 & -1 \end{vmatrix} + 4(-1)^{1+2}\begin{vmatrix} 1 & -2 \\ 2 & -1 \end{vmatrix} + 1(-1)^{1+3}\begin{vmatrix} 1 & 1 \\ 2 & 1 \end{vmatrix} = \begin{vmatrix} 1 & -2 \\ 1 & -1 \end{vmatrix} - 4\begin{vmatrix} 1 & -2 \\ 2 & -1 \end{vmatrix} + \begin{vmatrix} 1 & 1 \\ 2 & 1 \end{vmatrix} = -12.$$

228. (A) If every element in a column or row of a determinant is multiplied by a constant (here, row 1 is multiplied by 2), the new determinant is the constant times the original.

(B) In a given determinant, if any column or row is all zeros, the determinant is 0. Here, row 2 is all zeros.

(C) See question 228B. All entries here in column 1 are zero. Notice how much work is saved by using this theorem.

(D) If a constant multiple of a row or column is added to another row or column, the determinant's value is unchanged. Here we multiplied row 2 by −3 and added it to row 1.

(E) If two rows or columns in a given determinant are identical, then the determinant's value = 0. Here row 1 = row 3.

(F) If two rows or columns are interchanged in a given determinant, the resulting determinant is the negative of the original. Here, rows 1 and 2 were interchanged.

229. The product is $\begin{bmatrix} 3\cdot 3+(-4)(2) & 3\cdot 4+(-4)(3) \\ (-2)(3)+3\cdot 2 & (-2)(4)+3\cdot 3 \end{bmatrix} = \begin{bmatrix} 1 & 0 \\ 0 & 1 \end{bmatrix} = I$, the identity matrix.

Thus, they are inverses of each other. If $AB = I$, them $A = B^{-1}$.

230. We write the augmented matrix $\begin{bmatrix} 1 & 2 & 1 & 0 \\ 1 & 3 & 0 & 1 \end{bmatrix}$ and transform in to a matrix of the

form: $\begin{bmatrix} 1 & 0 \\ 1 & 1 \end{bmatrix} S$. When we do this, $S = A^{-1}$, where $A = \begin{bmatrix} 1 & 2 \\ 1 & 3 \end{bmatrix}$. We convert one matrix to

the other, using elementary row operations: (1) interchange two rows, (2) multiply a row

by $k \neq 0$, (3) add kR_i to R_j where R is a row in the matrix. In this case we have the following:

Add $-R_1$ to R_2 and replace R_2 by $R_2 - R_1$. Then $\begin{bmatrix} 1 & 2 & 1 & 0 \\ 1 & 3 & 0 & 1 \end{bmatrix} \sim \begin{bmatrix} 1 & 2 & 1 & 0 \\ 0 & 1 & -1 & 1 \end{bmatrix}$. Replace R_1

by $R_1 + (-2)R_2$. Then $\begin{bmatrix} 1 & 2 & 1 & 0 \\ 0 & 1 & -1 & 1 \end{bmatrix} \sim \begin{bmatrix} 1 & 0 & 3 & -2 \\ 0 & 1 & -1 & 1 \end{bmatrix}$. Thus, $M^{-1} = \begin{bmatrix} 3 & -2 \\ -1 & 1 \end{bmatrix}$. Check it!

$\begin{bmatrix} 3 & -2 \\ -1 & 1 \end{bmatrix}\begin{bmatrix} 1 & 2 \\ 1 & 3 \end{bmatrix} = \begin{bmatrix} 1 & 0 \\ 0 & 1 \end{bmatrix}$.

231. $\begin{bmatrix} 1 & -3 & 0 & 1 & 0 & 0 \\ 0 & 3 & 1 & 0 & 1 & 0 \\ 2 & -1 & 2 & 0 & 0 & 1 \end{bmatrix} \sim \begin{bmatrix} 1 & -3 & 0 & 1 & 0 & 0 \\ 0 & 3 & 1 & 0 & 1 & 0 \\ 0 & 5 & 2 & -2 & 0 & 1 \end{bmatrix}$ (We replaced R_3 by $R_3 - 2R_2$.)

$\sim \begin{bmatrix} 1 & -3 & 0 & 1 & 0 & 0 \\ 0 & 1 & \frac{1}{3} & 0 & \frac{1}{3} & 0 \\ 0 & 5 & 2 & -2 & 0 & 1 \end{bmatrix}$ (We replaced R_2 by $\frac{1}{3}R_2$.)

$\sim \begin{bmatrix} 1 & 0 & 1 & 1 & 1 & 0 \\ 0 & 1 & \frac{1}{3} & 0 & \frac{1}{3} & 0 \\ 0 & 0 & \frac{1}{3} & -2 & -\frac{5}{3} & 1 \end{bmatrix}$ (We replaced R_1 by $R_1 + 3R_2$, and R_3 by $R_3 - 5R_2$.)

$$\sim \begin{bmatrix} 1 & 0 & 1 \\ 0 & 1 & \frac{1}{3} \\ 0 & 0 & 1 \end{bmatrix} \begin{matrix} 1 & 1 & 0 \\ 0 & \frac{1}{3} & 0 \\ -6 & -5 & 3 \end{matrix} \quad \text{(We replaced } R_3 \text{, by } 3R_3.)$$

$$\sim \begin{bmatrix} 1 & 0 & 0 \\ 0 & 1 & 0 \\ 0 & 0 & 1 \end{bmatrix} \begin{matrix} 7 & 6 & -3 \\ 2 & 2 & -1 \\ -6 & -5 & 3 \end{matrix} \quad \begin{matrix} \text{(We replaced } R_1 \text{, by } R_1 - R_2, \\ R_2 \text{ by } R_2 - \frac{1}{3}.) \end{matrix}$$

Then $M^{-1} = \begin{bmatrix} 7 & 6 & -3 \\ 2 & 2 & -1 \\ -6 & -5 & 3 \end{bmatrix}$.

232. By Cramer's rule

$$x = \frac{\begin{vmatrix} k_1 & a_{12} \\ k_2 & a_{22} \end{vmatrix}}{D} \qquad y = \frac{\begin{vmatrix} a_{11} & k_1 \\ a_{21} & k_2 \end{vmatrix}}{D}$$

where $D = \begin{vmatrix} a_{11} & a_{12} \\ a_{21} & a_{22} \end{vmatrix} (\neq 0)$. Here $a_{11} = 1$, $a_{12} = 2$, $a_{21} = 3$, $a_{22} = -5$, $k_1 = 6$, and $k_2 = 10$.

$$x = \frac{\begin{vmatrix} 6 & 2 \\ 10 & -5 \end{vmatrix}}{D} \qquad y = \frac{\begin{vmatrix} 1 & 6 \\ 3 & 10 \end{vmatrix}}{D} \qquad D = \begin{vmatrix} 1 & 2 \\ 3 & -5 \end{vmatrix} = -5 - 6 = -11$$

Since $\begin{vmatrix} 6 & 2 \\ 10 & -5 \end{vmatrix} = -30 - 20 = -50$ and $\begin{vmatrix} 1 & 6 \\ 3 & 10 \end{vmatrix} = 10 - 18 = -8$, $x = \frac{-50}{-11} = \frac{50}{11}$, and $y = \frac{-8}{-11} = \frac{8}{11}$.

233. Given a system of equations

$$a_{11}x + a_{12}y + a_{13}z = k_1$$
$$a_{21}x + a_{22}y + a_{23}z = k_2$$
$$a_{31}x + a_{32}y + a_{33}z = k_3$$

then

$$x = \frac{\begin{vmatrix} k_1 & a_{12} & a_{13} \\ k_2 & a_{22} & a_{23} \\ k_3 & a_{32} & a_{33} \end{vmatrix}}{D} \qquad y = \frac{\begin{vmatrix} a_{11} & k_1 & a_{13} \\ a_{21} & k_2 & a_{23} \\ a_{31} & k_3 & a_{33} \end{vmatrix}}{D} \qquad z = \frac{\begin{vmatrix} a_{11} & a_{12} & k_1 \\ a_{21} & a_{22} & k_2 \\ a_{31} & a_{32} & k_3 \end{vmatrix}}{D}$$

$$D = \begin{vmatrix} a_{11} & \cdots & a_{13} \\ \vdots & & \vdots \\ a_{31} & \cdots & a_{33} \end{vmatrix} (\neq 0)$$

Here,
$$D = \begin{vmatrix} 1 & 1 & 0 \\ 0 & 2 & 1 \\ -1 & 0 & 1 \end{vmatrix} = 1$$

$$x = \dfrac{\begin{vmatrix} 0 & 1 & 0 \\ -5 & 2 & 1 \\ -3 & 0 & 1 \end{vmatrix}}{D} = \dfrac{2}{1} = 2 \quad y = \dfrac{\begin{vmatrix} 1 & 0 & 1 \\ 0 & -5 & 1 \\ -1 & -3 & 1 \end{vmatrix}}{D} = \dfrac{-2}{1} = -2 \quad z = \dfrac{\begin{vmatrix} 1 & 1 & 0 \\ 0 & 2 & -5 \\ -1 & 0 & -3 \end{vmatrix}}{D} = \dfrac{-1}{1} = -1$$

234. $y = 3x - 8$. Substituting, we get $26 = 3x^2 - (3x - 8)^2 = 3x^2 - (9x^2 - 48x + 64) = -6x^2 + 48x - 64 = 26$. Then $-6x^2 + 48x - 90 = 0$, $x^2 - 8x + 15 = 0$, $(x - 5)(x - 3) = 0$, and $x = 5$, $x = 3$. If $x = 5$, $y = 3(5) - 8 = 7$. If $x = 3$, $y = 3(3) - 8 = 1$. Check to see that these solutions are correct. They are!

235. Multiply (2) by 2 and add to (1):

$$x^2 + 2xy + y^2 = 49 \quad \text{or} \quad x + y = \pm 7$$

Multiply (2) by -2 and add to (1):

$$x^2 - 2xy + y^2 = 1 \quad \text{or} \quad x - y = \pm 1$$

Solve the system:

$$
\begin{aligned}
x + y &= 7 & x + y &= 7 \\
x - y &= 1 & x - y &= -1 \\
2x &= 8 & 2x &= 6 \\
x &= 4 & x &= 3 \\
y &= 7 - x = 3 & y &= 7 - x = 4
\end{aligned}
$$

$$
\begin{aligned}
x + y &= -7 & x + y &= -7 \\
x - y &= 1 & x - y &= -1 \\
2x &= -6 & 2x &= -8 \\
x &= -3 & x &= -4 \\
y &= -7 - x = -4 & y &= -7 - x = -3
\end{aligned}
$$

The solutions are $x = \pm 4$, $y = \pm 3$; $x = \pm 3$, $y = \pm 4$. See Figure A5.2.

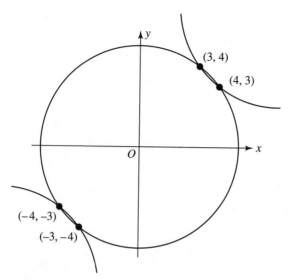

Figure A5.2

236. Let x and y be the numbers. Then $x - y = 2$ (differ by 2) and $x^2 - y^2 = 48$ (squares differ by 48). Thus, $x = y + 2$. Substituting for x, we get $(y + 2)^2 - y^2 = 48$, $y^2 + 4y + 4 - y^2 = 48$, $4y = 44$, and $y = 11$. If $y = 11$, $x = 13$.

237. Let r_1 and r_2 be the radii. Thus, $2\left(\dfrac{22}{7}\right)r_1 + 2\left(\dfrac{22}{7}\right)r_2 = 88$ and $\dfrac{22}{7}r_1^2 + \dfrac{22}{7}r_2^2 = \dfrac{2200}{7}$.
Then $44r_1 + 44r_2 = 616$, $r_1 + r_2 = 14$, $r_1 = 14 - r_2$. Also $r_1^2 + r_2^2 = 100$. Substituting for r_1 in this equation gives $(14 - r_2)^2 + r_2^2 = 100$, $196 + r_2^2 - 28r_2 + r_2^2 = 100$, $2r_2^2 - 28r + 96 = 0$, $r_2^2 - 14r + 48 = 0$, and $r = 6$ in, 8 in.

238. Let $x =$ cost per person and $y =$ number of people. Then $xy = 30$ and $x = \dfrac{30}{y}$. Thus, $x - 0.5 = \dfrac{30}{(y + 3)}$. (since the cost per person was reduced by 50¢ and the number of people was increased by 3). Then $\dfrac{30}{y} - 0.5 = \dfrac{30}{(y + 3)}$, and $y = 12$.

239. Let x and y be the numbers. Then $x^2 = 2y^2 + 16$ and $x^2 + y^2 = 208$. Thus, $x^2 = 208 - y^2 = 2y^2 + 16$, $2y^2 + 16 = 208 - y^2$, $3y^2 = 192$, $y^2 = 64$, and $y = \pm 8$. If $y = 8$, $x = \pm 12$. If $y = -8$, $x = \pm 12$. Thus the solutions are $x = 12$, $y = 8$; $x = 12$, $y = -8$; $x = -12$, $y = 8$; $x = -12$, $y = -8$.

240. See Figure A5.3(a). Let $x =$ width and $y =$ length. Then $d = \sqrt{x^2 + y^2}$, and $d' = \sqrt{(x + 11)^2 + (y - 7)^2}$ (See Figure A5.3(b)). If $d = d'$, then $\sqrt{x^2 + y^2} = \sqrt{(x + 11)^2 + (y - 7)^2}$. Since $\sqrt{x^2 + y^2} = 85$, $\sqrt{(x + 11)^2 + (y - 7)^2} = 85$. Then we have the equations

$(x + 11)^2 + (y − 7)^2 = 7225$ and $x^2 + y^2 = 7225$. Solving these two equations, we get $x = 40$ ft and $y = 75$ ft.

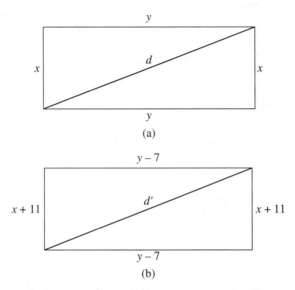

(a)

(b)

Figure A5.3

241. See Figure A5.4. We graph the line $2x − 3y = 6$. Then we find which "side" of the plane satisfies the inequality by testing a point. Dash the line $2x − 3y = 6$ since the inequality is <, not ≤. Test (0,0): $2(0) − 3(0) < 6$. Thus, the (0,0) side of the plane is shaded.

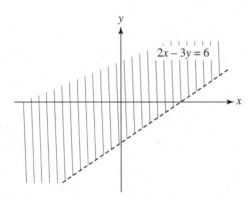

Figure A5.4

242. See Figure A5.5. We graph $2 \leq x < 2$ and $-1 < y \leq 6$ on the same set of axes and find where they intersect. Notice the crosshatched region: That is the solution set.

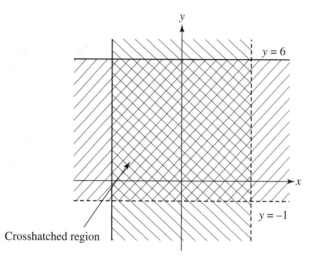

Crosshatched region

Figure A5.5

243. See Figure A5.6. Use $(0, 1)$ as a test. Is $1 > 0^2$? Yes! The crosshatched region is the solution set.

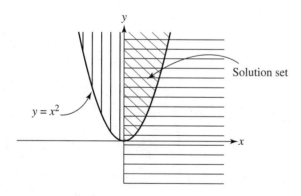

Figure A5.6

244. If the line l connects $P_1(x_1, y_1)$ and $P_2(x_2, y_2)$, then $x = x_1 + t(x_2 - x_1)$ and $y = y_1 + t(y_2 - y_1)$ for any $t \in \mathcal{R}$ represents (x, y) on l. In this case, $x = 2 + t(5 - 2)$ and $y = 3 + t(8 - 3)$, or $x = 2 + 3t$ and $y = 3 + 5t$. These last two equations represent (x, y) on line $\overline{P_1 P_2}$. If $0 \leq t \leq 1$, the segment $\overline{P_1 P_2}$ is represented.

245. If $f(t) = at + b$, $a \neq 0$ (where $a, b \in \mathcal{R}$), then if $c \leq t \leq d$, the extrema for f occur at c and d. Also if $a > 0$, the maximum is $f(d)$ and the minimum is $f(c)$. If $a < 0$, the maximum is $f(c)$ and the minimum is $f(d)$. Here $a = 2 > 0$; $c = 0$, $d = 4$. Thus, the maximum is $f(4) = 4(2) + 5 = 13$, and the minimum is $f(0) = 5$.

246. $f(x, y) = f[x_1 + t(x_2 - x_1), y_1 + t(y_2 - y_1)] = ax_1 + by_1 + c + [a(x_2 - x_1) + b(y_2 - y_1)]$ $t = g(t)$, where $P_1 = (x_1, y_1)$, $P_2 = (x_2, y_2)$ and $f(x, y) = ax + by + c$. Then $g(t) = ax_1 + by_1 + c + [a(x_2 - x_1) + b(y_2 - y_1)] t = g(t) = 3(2) + 2(1) + (-5) + [3(8 - 2) + 2(6 - 1)] t = 3 + (18 + 10)t = 3 + 28t$. Since $28 > 0$, the maximum is $g(1) = 31$ and the minimum is $g(0) = 3$.

Chapter 6: Exponential and Logarithmic Functions

247. See Figure A6.1. Since $2^x \cdot 2^{-x} = 1$, we suspect that something interesting will occur. Notice that one curve is the image of the other in the y axis.

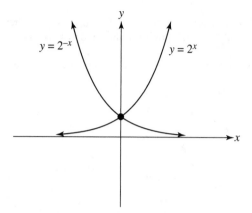

Figure A6.1

248. See Figure A6.2. Recall that e is the base for natural logarithms. In calculus you will learn that e is the number which $\left(1 + \dfrac{1}{b}\right)^n$ approaches as n gets arbitrarily large. Numerically, $e \approx 2.718$; it is an irrational number. Thus, when $x = 0$, $e^0 = 1$; as x increases, so does e^x.

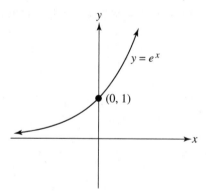

Figure A6.2

249. (A) $f(x+y) = a^{x+y} = a^x a^y = f(x)f(y)$.

(B) $f(x-y) = a^{x-y} = \dfrac{a^x}{a^y} = \dfrac{f(x)}{f(y)}$.

(C) Suppose that $m \neq n$. That $a^m \neq a^n$; f is one-to-one.

(D) $f(x) = a^x$; then $f(-x) = a^{-x} = \dfrac{1}{a^x} = \dfrac{1}{f(x)}$.

(E) $f(b+x) = a^{b+x} = a^b a^x = a^b f(x)$.

250. $A(n) = P(1 + i)^n$, where $A(n) =$ compound amount at the end of n interest periods, $i =$ rate per interest period, $P =$ amount invested. Here $n = 6$, $i = \frac{0.12}{2} = 0.06$. Then $A(6) = \$2000(1 + 0.06)^6 = \$2000(1.06)^6 \approx \$2000(1.42)$(rounded to two places on the calculator) $= \$2840$.

251. Since $A = P(1 + i)^n$, $P = A(1 + i)^{-n}$; here $A = \$5000$, $i = \frac{0.1}{1} = 0.1$, $n = 5.0$. Then $P = \$5000(1 + 0.1)^{-5} = \$5000(1.1)^{-5} \approx \$5000(0.62)$(rounded to the hundreds place) $= \$3100$.

252. $A = Pe^{rt} = \$3000(e^{5(0.1)}) = \$3000(e^{0.5}) \approx \$4946$ (rounded to the nearest dollar).

253. If $\log 100 = x$, then $10^x = 100 = 10^2$ and $x = 2$.

254. If $\log x = 2$, then $10^2 = x$ and $x = 100$.

255. $\log_x 81 = 4$. Then $x^4 = 81$, or $x = 3$ $(3^4 = 81)$.

256. Then $\log_e e^{x+2} = 7$. Thus, $e^7 = e^{x+2}$, $x + 2 = 7$, or $x = 5$.

257. Here, the base is e^2. Then $(e^2)^{10} = x$, or $x = e^{20}$.

258. (A) $f \circ g(x) = f(g(x)) = \log 10^x$. But $\log 10^x = x$. $f \circ g(x) = x$.
   (B) $g \circ f(x) = 10^{\log x} = x (x > 0)$.
   (C) $h \circ k(x) = \ln e^x = \log_e e^x = x$. Thus, $h \circ k(x) = x$.
   (D) $l(x) = x^2$; thus, $l(10) = 10^2 = 100$. Then $f \circ l(10) = f(l(10)) = \log 10^2 = 2$.
   (E) $h(x) = \ln x$; thus, $h(3) = \ln 3$. Then $l \circ h(3) = l(h(3)) = l(\ln 3) = (\ln 3)^2$.

259. See Figure A6.3. Since $y = \log_2 x$ and $y = 2^x$ are inverse functions, their graphs are mirror images about $y = x$.

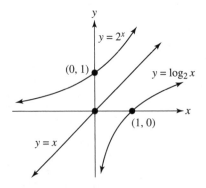

Figure A6.3

260. To prove this, we use the definition of one-to-one. If $\log_a x = \log_a y$ and $\log_a x = p$, then $\log_a y = p$, and $a^p = x$, $a^p = y$. Thus, $x = y$. $g(x)$ is one-to-one.

261. (A) We know $x + 1 \neq 0$; thus, the domain $= \{x \in \mathcal{R} | x > -1\}$, and the range $= \mathcal{R}$.
   (B) We must have $2x - 5 > 0$; thus, the domain $= \{x \in \mathcal{R} | x > \frac{5}{2}\}$, and the range $= \mathcal{R}$.

(C) Since $x^2 + 1 > 0 \;\forall x \in \mathcal{R}$, the domain $= \mathcal{R}$, and the range $= \mathcal{R}$.

(D) $x > 0$ for $\log_6 x$ to be defined. But $|\log_6 x| \geq 0 \;\forall x \in \mathcal{R}$; thus, the domain $= \{x \in \mathcal{R} \,|\, x > 0\}$, and the range $= \{y \in \mathcal{R} \,|\, y \geq 0\}$.

262. (A) $\log xy = \log x + \log y$; thus, $\log (251)(46)(18) = \log 251 + \log (46)(18) = \log 251 + \log 46 + \log 18$.

(B) $\log xy = \log x + \log y$, and $\log z^2 = 2 \log z$; thus, $\log (34)^2(2.7) = \log (34)^2 + \log (2.7) = 2 \log 34 + \log 2.7$.

(C) In general, $\log a^x = x \log a$; thus, $\log(24)^{\frac{1}{2}}(35)^3 = \log(24)^{\frac{1}{2}} + \log(35)^3 = \dfrac{1}{2}\log 24 + 3 \log 35$.

(D) $\log \dfrac{x}{y} = \log x - \log y$; thus, $\log \dfrac{(83)(41)}{29} = \log(83)(41) - \log 29 = (\log 83 + \log 41) - \log 29$.

(E) $\log a^n b^m = \log a^n + \log b^m = n \log a + m \log b$.

(F) $\log \sqrt[n]{a^{n-1} p} = \log(a^{n-1} p)^{\frac{1}{n}} = \left(\dfrac{1}{n}\right) \log a^{n-1} p = \left(\dfrac{1}{n}\right)(\log a^{n-1} + \log p) = \left(\dfrac{1}{n}\right) \times [(n-1)\log a + \log p] = \dfrac{n-1}{n}\log a + \dfrac{\log p}{n}$.

263. (A) $\log_2 (8)(16{,}384) = \log_2 8 + \log_2 16{,}384$. But $2^{14} = 16{,}384$. Thus, $\log_2 8 + \log_2 16{,}384 = 3 + 14 = 17$.

(B) $\log_2 (16{,}384)^{-2} = -2\log 16{,}384 = -2 \cdot 14 = -28$.

(C) $2^{16} = 65{,}536$. Thus, $\log_2 (65{,}536)^{\frac{1}{4}} = \dfrac{1}{4}\log_2 65{,}536 = \dfrac{1}{4} \cdot 16 = 4$.

264. (A) $2\log_b x = \log_b x^2$; thus, $2\log_b x - \log_b y = \log_b x^2 - \log_b y = \log_b \left(\dfrac{x^2}{y}\right)$.

(B) $3\log_b x + 2\log_b y - 4\log_b z = \log_b x^3 + \log_b y^2 - \log_b z^4 = \log_b \left(\dfrac{x^3 y^2}{z^4}\right)$.

(C) $\dfrac{1}{5}(2\log_b x + 3\log_b y) = \dfrac{2}{5}\log_b x + \dfrac{3}{5}\log_b y = \log_b x^{\frac{2}{5}} + \log_b y^{\frac{3}{5}} = \log_b x^{\frac{2}{5}} y^{\frac{3}{5}}$.

(D) $\log_b x - \log_b y = \log_b \left(\dfrac{x}{y}\right)$; thus $\dfrac{1}{3}\log_b \left(\dfrac{x}{y}\right) = \log_b \left(\dfrac{x}{y}\right)^{\frac{1}{3}}$.

265. (A) Using the property $\log_b x = \dfrac{\log_a x}{\log_a b}$, we want $b = 6$, $x = 7$, $a = 10$ (this is to be in terms of common logarithms). Then $\log_6 7 = \dfrac{\log 7}{\log 6}$ $(\log 7 = \log_{10} 7)$.

(B) $\log_{\frac{1}{3}} 30 = \log_b x$ where $b = \dfrac{1}{3}$, $x = 30$. Then $\log_b x = \log_{\frac{1}{3}} 30 = \dfrac{\log_a x}{\log_a b} = \dfrac{\log 30}{\log \frac{1}{3}} = \dfrac{\log(10 \cdot 3)}{(-\log 3)} = \dfrac{(\log 10 + \log 3)}{(-\log 3)} = \dfrac{(1 + \log 3)}{(-\log 3)}$.

(C) Here, e is *not* the base; 20 is. $\log_{20} e = \dfrac{\log_e}{\log 20} = \dfrac{\log e}{\log(10 \cdot 2)} = \dfrac{\log e}{\log 10 + \log 2} = \dfrac{\log e}{1 + \log 2}$.

(D) See question 265(C), $\log_{20} e^3 = 3\log_{20} e = 3\dfrac{\log e}{1 + \log 2} = \dfrac{3\log e}{1 + \log 2}$.

266. (A) This is true; $\log 15 \in \mathcal{R}$, and $a^0 = 1 \; \forall a \in \mathcal{R}$. Do not be tricked by the fact that the real number here is a logarithm. *All* real numbers can be written in logarithm form.

(B) True. Look at their graphs, or recall that the logarithm is the exponent. In $\log_a b = q$, a and b are the numbers which are restricted, not q.

(C) False. For example, if $x = 1$, then 1 is the logarithmic base, which is impossible.

(D) If $\log_x a = b$, then $x^b = a$, $\dfrac{1}{x^b} = \dfrac{1}{a}$, $\left(\dfrac{1}{x}\right)^b = \dfrac{1}{a}$, and $\log_{\frac{1}{x}} \dfrac{1}{a} = b$. The statement is true.

(E) False. $f(x) = \log_b x$ is decreasing for $0 < b < 1$. If $b = 10$, for example, $\log x_1 > \log x_2$ when $x_1 > x_2$.

(F) Clearly, this is a false statement. For all k, a, b, $\log_k ab = \log_k a + \log_k b$. When $k = e$, $a = e$, $b = e$, for example, $\ln(e \cdot e) = \ln e^2 = 2$, but in $e \ln e = 1 \cdot 1 = 1$; $2 \ne 1$.

(G) $h(ab) = \log_b ab = \log_b a + \log_b b = 1 + \log_b a$. $g(1 + \log_b a) = \log_a (1 + \log_b a)$. The statement is true.

(H) False. For this to be true, we would be saying that "the log of a sum is the sum of the logs." But this is false since, for example, $\log_2 8 = \log_2 2^3 = 3$, thus $\log_2 (4 + 4) = 3$, not $2 + 2$.

(I) $\log_a \left(\dfrac{1}{x}\right) = \log_a (x^{-1}) = -\log_a x$. True.

(J) $\log_b a = \dfrac{\log_k a}{\log_k b}$ for all log bases k. Let $k = a$; $\log_b a = \dfrac{\log_a a}{\log_a b} = \dfrac{1}{\log_a b}$. True.

267. Then $\log_b x = \log_b 4^{\frac{3}{2}} - \log_b 8^{\frac{2}{3}} + \log_b 2^2 = \log_b 8 - \log_b 4 + \log_b 4 = \log_b \left(\dfrac{8}{4} \cdot 4\right)$, or $x = 8$.

268. If $f(x) = 7^x$, then $f^{-1}(x) = \log_7 x$. Then the domain of f is $\mathcal{R}^+$, and the range (i.e., the f^{-1} values) is $\mathcal{R}$.

269. If $A = Pe^{rt}$, then $\ln A = \ln Pe^{rt} = \ln P + \ln e^{rt} = \ln P + rt \ln e = \ln P + rt$ (since $\ln e = 1$). Thus, $rt = \ln A - \ln P$, and $r = \dfrac{(\ln A - \ln P)}{t}$.

270. (A) If $y = 10^x$, then $x = \log y$. You can get that answer by remembering that $\log_a x$ and a^x are inverse functions, or by taking the log of both sides of the equation $y = 10^x$. Then $\log y = \log 10^x$ and $\log y = x$.

(B) $\log y = \log 3(10^{2x}) = \log 3 + \log 10^{2x} = \log 3 + 2x$. Then $2x = \log y - \log 3 = \log\left(\dfrac{y}{3}\right)$ and $x = \dfrac{1}{2}\log\left(\dfrac{y}{3}\right) = \log\left(\dfrac{y}{3}\right)^{\frac{1}{2}}$.

(C) Let $u = e^x$. (This is a common trick.) Then $e^{-x} = \dfrac{1}{u}$, so $2y = u + \dfrac{1}{u} = \dfrac{u^2 + 1}{u}$ and $u^2 - 2uy + 1 = 0$, a quadratic in u. Then $u = \dfrac{2y + \sqrt{4y^2 - 4}}{2} = y + \sqrt{y^2 - 1}$, and $e^x = y \pm \sqrt{y^2 - 1}$. Thus, $x = \ln(y \pm \sqrt{y^2 - 1})$.

271. $5^{x^2 - 3x} = 5^4$, $x^2 - 3x = 4$, $x^2 - 3x - 4 = 0$, $(x - 4)(x + 1) = 0$, or $x = 4, -1$. (Check these.)

272. $\log 195^x = \log 2.68$, $x \log 195 = \log 2.68$, or $x = \dfrac{\log 2.68}{\log 195} \approx \dfrac{0.4281}{2.2900} \approx 0.187$.

273. Then $\log_5 (x - 1)(x + 3) = 1$, $5^{\log_5 (x - 1)(x + 3)} = 5^1$, $(x - 1)(x + 3) = 5$, $x^2 + 2x - 3 = 5$, $x^2 + 2x - 8 = 0$, $(x + 4)(x - 2) = 0$, or $x = -4, 2$. It is crucial to check these answers since the domain of $\log x$ is so restricted. If $x = -4$, $\log_5 (-5) + \log_5 (-1)$ is undefined and $x = -4$ is extraneous. If $x = 2$, $\log_5 1 + \log_5 5 = 0 + 1 = 1$. Thus, the solution is $x = 2$.

274. Then $3x + 4 = 5x - 6$, $2x = 10$, or $x = 5$. Check this solution.

275. If $2 \ln 5x = 3 \ln x$, then $\ln (5x)^2 = \ln x^3$. Then $(5x)^2 = x^3$, $25x^2 = x^3$, $x^3 - 25x^2 = 0$, $x^2(x - 25) = 0$, or $x = 0, 25$. But $x = 0$ is extraneous, since $5x = 0$ and $\ln 0$ is undefined.

276. If $\log x = \ln e$, then $10^{\log x} = 10^{\ln e}$, or $x = 10^{\ln e}$. But $\ln e = 1$, so $x = 10$.

Chapter 7: Trigonometric Functions

277. See Figure A7.1. $180° + 45° = 225°$.

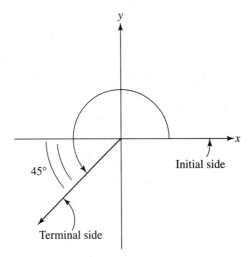

Figure A7.1

278. See Figure A7.2. $420° = 360° + 60°$.

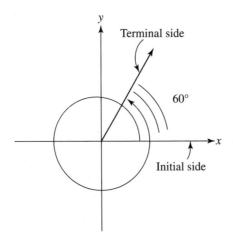

Figure A7.2

279. See Figure A7.3. $-\dfrac{\pi}{2} < -\dfrac{\pi}{3} < 0.$

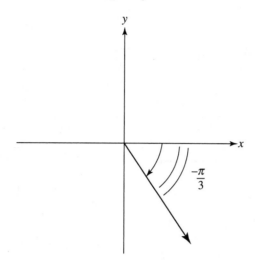

Figure A7.3

280. $114° - 81° = 33°$; $29' - 4' = 25'$; $46'' - 11'' = 35''$. So the answer is $33°25'35''$.

281. $40.25° = 40° + 0.25° = 40° + 0.25 \times 60'$ (since $60' = 1°$) $= 40° + 15' = 40°15' = 40°15'0''$.

282. $27°15'25'' = 27° + \left(\dfrac{15}{60}\right)° + \left[\dfrac{25}{(60)^2}\right]° = 27° + 0.25° + 0.00694\overline{4}° = 27.25694\overline{4}°.$

283. **(A)** 2π rad $= 360°$. Then $\dfrac{\pi}{4}$ rad $= 360° \times \dfrac{\left(\dfrac{\pi}{4}\right)}{(2\pi)} = 360° \times \dfrac{1}{8} = 45°.$

(B) Alternatively, $\pi = 180°$ and $\dfrac{\pi}{4} = \dfrac{180°}{4} = 45°.$

284. $-\dfrac{\pi}{6} = 180° \times (-\tfrac{1}{6}) = -30°.$

285. $180° = \pi$ rad. $45° = \pi \times \dfrac{45}{180} = \pi \times \dfrac{1}{4} = \dfrac{\pi}{4}$ rad.

286. $-135° = \pi \times \left(-\dfrac{135}{180}\right) = \pi \times (-0.75) = -0.75\pi = -\dfrac{3\pi}{4}$ rad.

287. $\theta = \dfrac{s}{R}$ (in radians) $= \dfrac{24 \text{ cm}}{4 \text{ cm}} = 6$ rad.

288. See Figure A7.4.
    (1) $360° + (−155°) = 205.$
    (2) $2 \times 360° + (−155°) = 565°.$
    (3) $3 \times 360° + (−155°) = 925°.$
    (4) $15 \times 360° + (−155°) = 5245°.$
    (5) $5245° − (35 \times 360°) = 5245° − 12{,}600° = −7355°.$

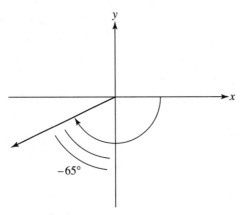

Figure A7.4

289. During 20 min, the hand moves through an angle $\theta = 120° = \dfrac{2\pi}{3}$ rad, and the tip of the hand moves a distance $s = r\theta = 12\left(\dfrac{2\pi}{3}\right) = 8\pi$ in $= 25.1$ in.

290. We are required to find the radius of a circle on which a central angle $\theta = 25° = \dfrac{5\pi}{36}$ rad intercepts an arc of 120 ft. Then

$$r = \frac{s}{\theta} = \frac{120}{\frac{5\pi}{36}} = \frac{864}{\pi} \text{ ft} = 275 \text{ ft}$$

291. Since $36° = \dfrac{\pi}{5}$ rad, $s = r\theta = 3960\left(\dfrac{\pi}{5}\right) = 2488$ mi.

292. Since $\sin\theta = \dfrac{y}{r}$ and $\cos\theta = \dfrac{x}{r}$, both x and y are negative. (Recall that r is always positive.) Thus, θ is a third-quadrant angle.

293. Since $\sec\theta$ is negative, x is negative; since $\tan\theta$ is negative, y is positive. Thus, θ is a second-quadrant angle.

294. Since $\sin\theta$ is positive, y is positive. Then x may be positive or negative, and θ is a first- or second-quadrant angle.

295. Since $\tan\theta$ is negative, either y is positive and x is negative or y is negative and x is positive. Thus, θ may be a second- or fourth-quadrant angle.

296. Since $\sin \theta$ is negative, θ is in quadrant III or IV.

    (A) In quadrant III: Take $y = m$, $r = n$, $x = -\sqrt{n^2 - m^2}$; then

$$\cos \theta = \frac{x}{r} = \frac{-\sqrt{n^2 - m^2}}{n} \quad \text{and} \quad \tan \theta = \frac{y}{x} = \frac{-m}{\sqrt{n^2 - m^2}}.$$

    (B) In quadrant IV: Take $y = m$, $r = n$, $x = +\sqrt{n^2 - m^2}$; then

$$\cos \theta = \frac{x}{r} = \frac{\sqrt{n^2 - m^2}}{n} \quad \text{and} \quad \tan \theta = \frac{y}{x} = \frac{m}{\sqrt{n^2 - m^2}}.$$

297. $0 + 2(1) + 3(1) + 4(0) + 5(1) + 6(1) = 16.$

298. See Figure A7.5. Here the point $(0, 1)$ is chosen on the terminal ray of the angle. From the equation $\sin \theta = \dfrac{b}{R}$ for any angle θ where b is the ordinate of (a, b) on the terminal ray, we have $\sin 90° = \frac{1}{1} = 1$. [Clearly, $R = 1$, the distance from $(0, 0)$ to $(0, 1)$.]

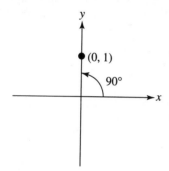

Figure A7.5

299. See Figure A7.6. For convenience, choose $(1, 1)$ on the terminal ray. Then $\sin \theta = \dfrac{b}{R} = \dfrac{1}{\sqrt{2}} = \dfrac{\sqrt{2}}{2}$ [$\sqrt{2}$ = distance from $(0, 0)$ to $(1, 1)$].

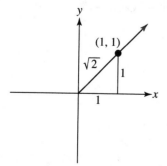

Figure A7.6

300. See Figure A7.7. $\cos \theta = \dfrac{a}{R} = \dfrac{1}{2}$. We choose $(1, \sqrt{3})$ on the 60° ray for convenience.

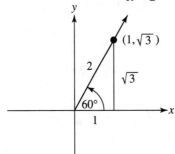

Figure A7.7

301. (a) We illustrate the 120° angle in Figure A7.8. The reference angle is found by dropping a perpendicular line to the x axis. Here $\theta = 180° - 120° = 60°$. Choose a convenient P on the terminal ray of 120°; $(-1, \sqrt{3})$ is convenient. Then $R = 2$, $b = \sqrt{3}$, and $\sin 120° = \dfrac{b}{R} = \dfrac{\sqrt{3}}{2}$.

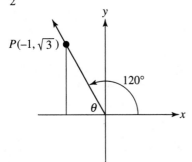

Figure A7.8

(b) Notice that (1) the reference angle for 120° is 60° (see Figure A7.9), (2) $\sin \theta > 0$ in quadrant II, and (3) $\sin 60° = \dfrac{\sqrt{3}}{2}$. Thus, $\sin 120° = +\sin 60° = \dfrac{\sqrt{3}}{2}$.

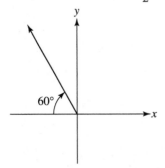

Figure A7.9

302. See Figure A7.10.

    (a) $\cos 135° = \dfrac{a}{R} = -\dfrac{1}{\sqrt{2}} = -\dfrac{\sqrt{2}}{2}$.

    (b) $\cos 135° = -\cos 45°$ ($\cos \theta < 0$ in quadrant II and $45°$ = reference angle) $= -\dfrac{\sqrt{2}}{2}$.

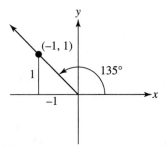

Figure A7.10

303. See Figure A7.11. $\sin\left(\dfrac{3\pi}{4}\right) = \sin\left(\dfrac{\pi}{4}\right) = \dfrac{\sqrt{2}}{2}$. $\left[\text{Recall}: \sin\left(\dfrac{\pi}{4}\right) = \dfrac{b}{R} = \dfrac{1}{\sqrt{2}} = \dfrac{\sqrt{2}}{2}.\right]$

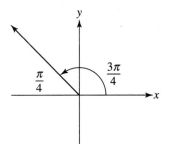

Figure A7.11

304. See Figure A7.12.

    (a) $\sec\left(\dfrac{4\pi}{3}\right) = -\sec\left(\dfrac{\pi}{3}\right) = -\dfrac{1}{\cos\left(\dfrac{\pi}{3}\right)} = -\dfrac{1}{\frac{1}{2}} = -2.$

    (b) $\sec\left(\dfrac{4\pi}{3}\right) = \dfrac{R}{a} = \dfrac{2}{-1} = -2.$

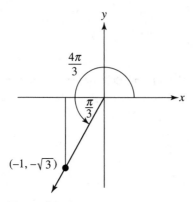

Figure A7.12

305. See Figure A7.13. sin (−960°) = −sin 960° [sin (−θ) = −sin θ] = −sin (720° + 240°) = −(−sin 60°) = sin 60° = $\dfrac{\sqrt{3}}{2}$.

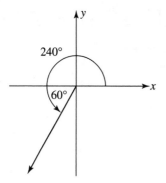

Figure A7.13

306. sin 315° = −sin 45° = −$\dfrac{\sqrt{2}}{2}$, and cos 315° = cos 45° = $\dfrac{\sqrt{2}}{2}$. Then sin² 315° + cos² 315° = $\left(-\dfrac{\sqrt{2}}{2}\right)^2 + \left(\dfrac{\sqrt{2}}{2}\right)^2 = \dfrac{1}{2} + \dfrac{1}{2} = 1$.

307. $\cos\left(\dfrac{\pi}{4}\right) = \dfrac{\sqrt{2}}{2}$, $\cos\left(\dfrac{\pi}{2}\right) = 0$, $\sin\left(\dfrac{\pi}{4}\right) = \dfrac{\sqrt{2}}{2}$, and $\sin\left(\dfrac{\pi}{2}\right) = 1$. Thus, $\cos\left(\dfrac{\pi}{4}\right)\cos\left(\dfrac{\pi}{2}\right) - \sin\left(\dfrac{\pi}{4}\right)\sin\left(\dfrac{\pi}{2}\right) = \dfrac{\sqrt{2}}{2} \cdot 0 - \dfrac{\sqrt{2}}{2} \cdot 1 = -\dfrac{\sqrt{2}}{2}$.

308. $\sin\theta = \dfrac{3}{5} = \dfrac{b}{R}$. Then from $a^2 + b^2 = R^2$ we have $a^2 + 9 = 25$, or $a = \pm 4$. But $\cos\theta < 0$,

so $a = -4$. Then $\cos\theta = \dfrac{a}{R} = -\dfrac{4}{5}$; $\tan\theta = \dfrac{b}{a} = -\dfrac{3}{4}$; $\cot\theta = -\dfrac{4}{3}$; $\csc\theta = \dfrac{1}{\sin\theta} = \dfrac{5}{3}$; and $\sec\theta =$

$\dfrac{1}{\cos\theta} = -\dfrac{5}{4}$.

309. $\tan\theta = -\dfrac{4}{3} = \dfrac{b}{a}$. Then $b = 4$, $a = -3$ or $b = -4$, $a = 3$. But $\sin\theta = \dfrac{b}{R}$ and $\sin\theta < 0$.

Thus, $b < 0$. Then $b = -4$, $a = 3$, and $R^2 = a^2 + b^2 = 25$, or $R = 5$ ($R > 0$ always). Then

$\cot\theta = -\dfrac{3}{4}$; $\sin\theta = \dfrac{b}{R} = -\dfrac{4}{5}$; $\csc\theta = \dfrac{1}{\sin\theta} = -\dfrac{5}{4}$; $\cos\theta = \dfrac{a}{R} = \dfrac{3}{5}$; and $\sec\theta = -\dfrac{5}{3}$.

310. If $y = \sin^{-1}\dfrac{1}{2}$, then $\sin y = \dfrac{1}{2}$ (by definition of the $\sin^{-1}x$, function). Then $y = \dfrac{\pi}{6}$ (30°).

Remember: If $y = \sin^{-1}x$, then $-\dfrac{\pi}{2} \le y \le \dfrac{\pi}{2}$. If $y = \cos^{-1}x$, then $0 \le y \le \pi$. If $y = \tan^{-1}x$,

then $-\dfrac{\pi}{2} < y < \dfrac{\pi}{2}$.

311. If $y = \arctan\sqrt{3}$, then $\tan y = \sqrt{3}$ $\left(-\dfrac{\pi}{2} < y < \dfrac{\pi}{2}\right)$ and $y = 60°$. (Did you forget that

$\tan 60° = \sqrt{3}$? If so, recall that $\sin 60° = \dfrac{\sqrt{3}}{2}$, $\cos 60° = \dfrac{1}{2}$, and $\tan 60° = \dfrac{\sin 60°}{\cos 60°}$.$\Big)$

312. $\arccos 0.5 = y$ means $\cos y = \dfrac{1}{2}$, or $y = 60°$. Then $\tan (\arccos 0.5) = \tan 60° =$

$\dfrac{\sin 60°}{\cos 60°} = \dfrac{\frac{\sqrt{3}}{2}}{\frac{1}{2}} = \sqrt{3}$.

313. $y = 4\sin(2x + \pi)$ which is of the form $y = A\sin(Bx + C)$ where $A = 4$, $B = 2$. Then

the period $= \dfrac{2\pi}{B} = \dfrac{2\pi}{2} = \pi$.

314. Here, $f(x) = A\tan(Bx + C)$ where $B = 2$. Then the period $= \dfrac{\pi}{B} = \dfrac{\pi}{2}$.

315. If $y = A\sin(Bx + C)$, $B > 0$, then the phase shift $= \left|\dfrac{C}{B}\right|$ to the right if $\dfrac{C}{B} < 0$; $\dfrac{C}{B}$ to

the left if $\dfrac{C}{B} > 0$. Here $A = 1$, $B = 1$, $C = \pi$, and $\dfrac{C}{B} = \dfrac{\pi}{1} > 0$; thus, the phase shift $= \pi$ units

to the left.

316. Here $B = 2$, and $C = -\dfrac{\pi}{7}$. Then $\dfrac{C}{B} = -\dfrac{\pi}{14} < 0$. Phase shift $= \left|\dfrac{C}{B}\right| = \dfrac{\pi}{14}$ to the right.

317. See Figure A7.14. Period of $\sin 2x = \dfrac{2\pi}{2} = \pi$. Both functions have an amplitude $= 1$.

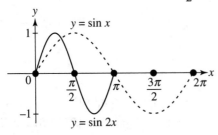

Figure A7.14

318. See Figure A7.15. Both functions have a period $= 2\pi$; $2 \sin x$ has amplitude $= 2$.

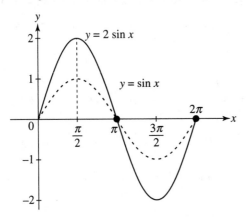

Figure A7.15

319. See Figure A7.16. The phase shift is $\dfrac{C}{B} = \dfrac{\left(\frac{\pi}{3}\right)}{1} = \dfrac{\pi}{3}$ units to the left $\left(\dfrac{C}{B} > 0\right)$.

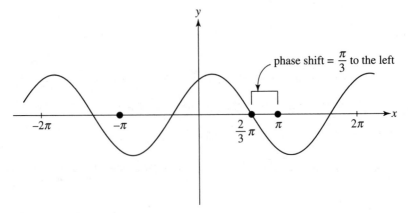

Figure A7.16

320. See Figure A7.17. Period $= \dfrac{\pi}{2}$.

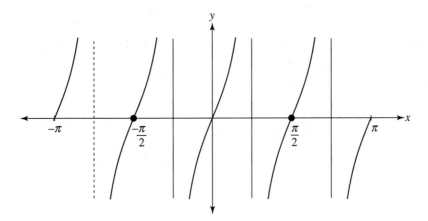

Figure A7.17

321. See Figure A7.18. This is the inverse of $y = \sin x \left(-\dfrac{\pi}{2} \le x \le \dfrac{\pi}{2} \right)$. Thus, if $y = \arcsin x$, then $\sin y = x$ where $-\dfrac{\pi}{2} \le y \le \dfrac{\pi}{2}$, $x \in [-1, 1]$.

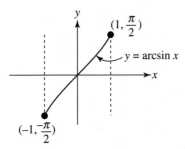

Figure A7.18

Chapter 8: Trigonometric Equations and Inequalities

322. Ask yourself, which side of the equation is more complicated? Begin working there. Continue to simplify that side of the equation until it looks exactly like the other side of the equation. $\sin x \, (\csc x - \sin x) = \sin x \csc x - \sin^2 x = 1 - \sin^2 x$ (since $\sin x = \dfrac{1}{\csc x}$) $= \cos^2 x$ (since $\sin^2 x + \cos^2 x = 1$), and the identity is established.

323. $\cos x \, (\sec x - \cos x) = \cos x \sec x - \cos^2 x = 1 - \cos^2 x = \sin^2 x$ (since $\sec x \cos x = 1$, $1 - \cos^2 x = \sin^2 x$).

324.
$$\frac{\sin x + \cos x}{\tan x} = \frac{\sin x}{\tan x} + \frac{\cos x}{\tan x}$$
$$= \frac{\sin x}{\frac{\sin x}{\cos x}} + \frac{\cos x}{\frac{\sin x}{\cos x}} \qquad \left(\text{since } \tan x = \frac{\sin x}{\cos x}\right)$$
$$= \frac{\sin x \cos x}{\sin x} + \frac{\cos x \cos x}{\sin x} = \cos x + \frac{\cos^2 x}{\sin x}$$

325.
$$\frac{\sin^2 x}{1 - \sin^2 x} + \frac{\cos^2 x}{1 - \cos^2 x} = \frac{\sin^2 x}{\cos^2 x} + \frac{\cos^2 x}{\sin^2 x} = \tan^2 x + \cot^2 x \left(\text{since } \frac{\sin x}{\cos x} = \tan x\right)$$
$$= \tan^2 x + \frac{1}{\tan^2 x} = \frac{\tan^4 x + 1}{\tan^2 x}$$

326. $(1 - \sin x)(1 + \sin x) + \sin^2 x = (1 - \sin^2 x) + \sin^2 x = \cos^2 x + \sin^2 x = 1$ or, more simply, $(1 - \sin^2 x) + \sin^2 x = 1 + 0 = 1$.

327. This is the same as question 328. Here we begin with the right-hand side. Sometimes both sides look fairly complicated, and it is hard to decide where to begin.
$$\tan x + \csc^2 x - \cot^2 x = \tan x + 1 = \frac{1}{\cot x} + \frac{\cot x}{\cot x} = \frac{1 + \cot x}{\cot x}$$

328.
$$\frac{1 + \cot x}{\cot x} = \frac{1}{\cot x} + \frac{\cot x}{\cot x} = \tan x + 1$$
$$= \tan x + (\csc^2 x - \cot^2 x) \qquad (\text{since } \csc^2 x - \cot^2 x = 1)$$

329.
$$(\sin x + \cos x)^4 = (\sin x + \cos x)^2 (\sin x + \cos x)^2$$
$$= (\sin^2 x + 2 \sin x \cos x + \cos^2 x)(\sin^2 x + 2 \sin x \cos x + \cos^2 x)$$
$$= (1 + 2 \sin x \cos x)(1 + 2 \sin x \cos x) = 1 + 4 \sin x \cos x + 4(\sin x \cos x)^2$$

330.
$$\frac{\cos^2\theta - \sin^2\theta}{\sin\theta \cos\theta} = \frac{\cos^2\theta}{\sin\theta \cos\theta} - \frac{\sin^2\theta}{\sin\theta \cos\theta}$$
$$= \frac{\cos\theta \cos\theta}{\sin\theta \cos\theta} - \frac{\sin\theta \sin\theta}{\sin\theta \cos\theta} = \cot\theta - \tan\theta$$

Note the first line of the equations. There were many algebraic techniques that seemed applicable here. The minus sign on both sides convinced us to use this technique. Do not be afraid to abort an attempt that appears futile!

331. $\cos^2\theta = 1 - \sin^2\theta$ and $\cos\theta = \pm\sqrt{1-\sin^2\theta}$

$$\tan\theta = \frac{\sin\theta}{\cos\theta} = \frac{\sin\theta}{\pm\sqrt{1-\sin^2\theta}} \qquad \cot\theta = \frac{1}{\tan\theta} = \frac{\pm\sqrt{1-\sin^2\theta}}{\sin\theta}$$

$$\sec\theta = \frac{1}{\cos\theta} = \frac{1}{\pm\sqrt{1-\sin^2\theta}} \qquad \csc\theta = \frac{1}{\sin\theta}$$

Note that $\cos\theta = \pm\sqrt{1-\sin^2\theta}$. Writing $\cos\theta = \sqrt{1-\sin^2\theta}$ limits angle θ to those quadrants (first and fourth) in which the cosine is positive.

332. $\sec^2\theta = 1 + \tan^2\theta$ and $\sec\theta = \pm\sqrt{1+\tan^2\theta}$, $\cos\theta = \frac{1}{\sec\theta} = \frac{1}{\pm\sqrt{1+\tan^2\theta}}$, $\frac{\sin\theta}{\cos\theta} = \tan\theta$

and $\sin\theta = \tan\theta\cos\theta = \tan\theta\frac{1}{\pm\sqrt{1+\tan^2\theta}} = \frac{\tan\theta}{\pm\sqrt{1+\tan^2\theta}}$, $\csc\theta = \frac{1}{\sin\theta} = \frac{\pm\sqrt{1+\tan^2\theta}}{\tan\theta}$,

$\cot\theta = \frac{1}{\tan\theta}$.

333. From $\cos^2\theta = 1 - \sin^2\theta$, $\cos\theta = \pm\sqrt{1-\sin^2\theta} = \pm\sqrt{1-\left(\frac{3}{5}\right)^2} = \pm\sqrt{\frac{16}{25}} = \pm\frac{4}{5}$.

Now $\sin\theta$ and $\cos\theta$ are both positive when θ is a first-quadrant angle while $\sin\theta = +$ and $\cos\theta = -$ when θ is a second-quadrant angle. Thus,

First quadrant

$\sin\theta = \frac{3}{5}$ $\qquad$ $\cot\theta = \frac{4}{3}$

$\cos\theta = \frac{4}{5}$ $\qquad$ $\sec\theta = \frac{5}{4}$

$\tan\theta = \frac{\frac{3}{5}}{\frac{4}{5}} = \frac{3}{4}$ $\qquad$ $\csc\theta = \frac{5}{3}$

Second quadrant

$\sin\theta = \frac{3}{5}$ $\qquad$ $\cot\theta = -\frac{4}{3}$

$\cos\theta = -\frac{4}{5}$ $\qquad$ $\sec\theta = -\frac{5}{4}$

$\tan\theta = -\frac{3}{4}$ $\qquad$ $\csc\theta = \frac{5}{3}$

334. Since $\tan\theta = -$, θ is either a second- or fourth-quadrant angle.

Second quadrant

$\tan\theta = -\frac{5}{12}$

$\cot\theta = \frac{1}{\tan\theta} = -\frac{12}{5}$

$\sec\theta = -\sqrt{1+\tan^2\theta} = -\frac{13}{12}$

$\cos\theta = \frac{1}{\sec\theta} = -\frac{12}{13}$

$\csc\theta = \sqrt{1+\cot^2\theta} = \frac{13}{5}$

$\sin\theta = \frac{1}{\csc} = \frac{5}{13}$

Fourth quadrant

$\tan\theta = -\frac{5}{12}$

$\cot\theta = -\frac{12}{5}$

$\sec\theta = \frac{13}{12}$

$\cos\theta = \frac{12}{13}$

$\csc\theta = -\frac{13}{5}$

$\sin\theta = -\frac{5}{13}$

335. (A) $(\sin \theta - \cos \theta)(\sin \theta + \cos \theta) = \sin^2 \theta - \cos^2 \theta$

(B) $(\sin A + \cos A)^2 = \sin^2 A + 2 \sin A \cos A + \cos^2 A$

(C) $(\sin x + \cos y)(\sin y - \cos x) = \sin x \sin y - \sin x \cos x + \sin y \cos y - \cos x \cos y$

(D) $(\tan^2 A - \cot A)^2 = \tan^4 A - 2 \tan^2 A \cot A + \cot^2 A$

(E) $1 + \dfrac{\cos\theta}{\sin\theta} = \dfrac{\sin\theta + \cos \theta}{\sin\theta}$

(F) $1 - \dfrac{\sin \theta}{\cos\theta} + \dfrac{2}{\cos^2 \theta} = \dfrac{\cos^2 \theta - \sin\theta \cos\theta + 2}{\cos^2 \theta}$

336. (A) $\sin^2 \theta - \sin \theta \cos \theta = \sin \theta(\sin \theta - \cos \theta)$

(B) $\sin^2 \theta + \sin^2 \theta \cos^2 \theta = \sin^2 \theta (1 + \cos^2 \theta)$

(C) $\sin^2 \theta + \sin \theta \sec \theta - 6 \sec^2 \theta = (\sin \theta + 3 \sec \theta)(\sin \theta - 2 \sec \theta)$

(D) $\sin^3 \theta \cos^2 \theta - \sin^2 \theta \cos^3 \theta + \sin \theta \cos^2 \theta = \sin \theta \cos^2 \theta (\sin^2 \theta - \sin \theta \cos \theta + 1)$

(E) $\sin^4 \theta - \cos^4 \theta = (\sin^2 \theta + \cos^2 \theta)(\sin^2 \theta - \cos^2 \theta)$
$= (\sin^2 \theta + \cos^2 \theta)(\sin \theta - \cos \theta)(\sin \theta + \cos \theta)$

337. (A) $\sec \theta - \sec \theta \sin^2 \theta = \sec \theta (1 - \sin^2 \theta) = \sec \theta \cos^2 \theta = \dfrac{1}{\cos\theta} \cos^2 \theta = \cos \theta$

(B) $\sin \theta \sec \theta \cot \theta = \sin \theta \cdot \dfrac{1}{\cos \theta} \cdot \dfrac{\cos\theta}{\sin\theta} = \dfrac{\sin \theta \cos \theta}{\cos \theta \sin \theta} = 1$

(C) $\sin^2 \theta(1 + \cot^2 \theta) = \sin^2 \theta \csc^2 \theta = \sin^2 \theta \cdot \dfrac{1}{\sin^2 \theta} = 1$

(D) $\sin^2 \theta \sec^2 \theta - \sec^2 \theta = (\sin^2 \theta - 1) \sec^2 \theta = - \cos^2 \theta \sec^2 \theta = - \cos^2 \theta \cdot \dfrac{1}{\cos^2 \theta} = -1$

(E) $(\sin \theta + \cos \theta)^2 + (\sin \theta - \cos \theta)^2 = \sin^2 \theta + 2 \sin \theta \cos \theta + \cos^2 \theta + \sin^2 \theta - 2 \sin \theta \cos \theta + \cos^2 \theta = 2(\sin^2 \theta + \cos^2 \theta) = 2$

(F) $\tan^2 \theta \cos^2 \theta + \cot^2 \theta \sin^2 \theta = \dfrac{\sin^2 \theta}{\cos^2 \theta} \cos^2 \theta + \dfrac{\cos^2 \theta}{\sin^2 \theta} \sin^2 \theta = \sin^2 \theta + \cos^2 \theta = 1$

338. Here we use the identity $\tan (x \pm y) = \dfrac{\tan x \pm \tan y}{1 \mp \tan x \tan y}$. Thus, $\tan 75° = \tan (45° + 30°) =$

$\dfrac{\tan 45° + \tan 30°}{1 - \tan 45° \tan 30°} = \dfrac{1 + \frac{1}{\sqrt 3}}{1 - 1\left(\frac{1}{\sqrt 3}\right)} = \dfrac{\frac{(\sqrt 3 + 1)}{\sqrt 3}}{\frac{(\sqrt 3 - 1)}{\sqrt 3}} = \dfrac{\sqrt 3 + 1}{\sqrt 3 - 1}$. If you forgot what $\tan 30°$ is, use

$\dfrac{\sin 30°}{\cos 30°}$.

339. $\tan \dfrac{7\pi}{12} = \tan \left(\dfrac{\pi}{3} + \dfrac{\pi}{4}\right) = \dfrac{\tan(\frac{\pi}{3}) + \tan(\frac{\pi}{4})}{1 - \tan(\frac{\pi}{3})\tan(\frac{\pi}{4})} = \dfrac{\sqrt 3 + 1}{1 - \sqrt 3 \cdot 1} = \dfrac{\sqrt 3 + 1}{1 - \sqrt 3}$.

340. $\dfrac{\tan 80° + \tan 40°}{1 - \tan 80° \tan 40°} = \tan (80° + 40°) = \tan 120° = - \tan 60°(\text{quadrant II}) = -3$.

341. $\dfrac{\sin(\theta + h) - \sin\theta}{h} = \dfrac{\sin \theta \cos h + \cos\theta \sin h - \sin\theta}{h}$

$= \dfrac{\sin \theta \cos h}{h} + \dfrac{\cos\theta \sin h}{h} - \dfrac{\sin\theta}{h}$

$$= \frac{\sin\theta\cos h}{h} - \frac{\sin\theta}{h} + \frac{\cos\theta\sin h}{h}$$

$$= \frac{\sin\theta(\cos h - 1)}{h} + \frac{\cos\theta\sin h}{h}$$

$$= \frac{-\sin\theta(1 - \cos h)}{h} + \frac{\cos\theta\sin h}{h}$$

342. See Figure A8.1. $\sin(x+y) = \sin x \cos y + \cos x \sin y$. Let $x = \cos^{-1}\left(-\frac{4}{5}\right)$ and $y = \sin^{-1}\left(-\frac{3}{5}\right)$.

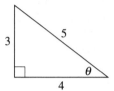

Figure A8.1

We get $\underbrace{\sin\left[\cos^{-1}\left(-\frac{4}{5}\right)\right]}_{+\frac{3}{5}\text{ (quad. II)}} \underbrace{\cos\left[\sin^{-1}\left(-\frac{3}{5}\right)\right]}_{+\frac{4}{5}\text{ (quad. IV)}} + \underbrace{\cos\left[\cos^{-1}\left(-\frac{4}{5}\right)\right]}_{\text{quad. II}} \underbrace{\sin\left[\sin^{-1}\left(-\frac{3}{5}\right)\right]}_{\text{quad. IV}}$

$$= \left(+\frac{3}{5}\right)\cdot\left(+\frac{4}{5}\right) + \left(-\frac{4}{5}\right)\left(-\frac{3}{5}\right) = +\frac{12}{25} + \frac{12}{25} = \frac{24}{25}$$

343. $\sin(x+y) = \sin x \cos y + \cos x \sin y$. Thus, letting $x = y$ in the above equation gives $\sin(x+x) = \sin x \cos x + \cos x \sin x = 2 \sin x \cos x$, and the proof is complete.

344. Letting $x = y$, $\cos(x+y) = \cos x \cos y - \sin x \sin y = \cos x \cos x - \sin x \sin x = \cos^2 x - \sin^2 x = (1 - \sin^2 x) - \sin^2 x = 1 - 2\sin^2 x$. From $\cos^2 x - \sin^2 x$ we also have $= \cos^2 x - (1 - \cos^2 x) = 2\cos^2 x - 1$, and the proof is complete.

345. $\sin 22.5° = \sin\left(\frac{45°}{2}\right)$. Since $\sin\frac{x}{2} = \pm\sqrt{\frac{1-\cos x}{2}}$, $\sin\frac{45°}{2} = \pm\sqrt{\frac{1-\cos 45°}{2}}$. We reject the minus answer since this is a quadrant I angle. Thus, the answer is $= \sqrt{\frac{1-\frac{1}{\sqrt{2}}}{2}}$.

346. $\tan\frac{x}{2} = \frac{1-\cos x}{\sin x}$. Then $\tan 165° = \frac{1-\cos 330°}{\sin 330°} = \frac{1-\sqrt{\frac{3}{2}}}{-\frac{1}{2}}\cdot\frac{2}{2} = \frac{2-\sqrt{3}}{-1} = \sqrt{3} - 2$.

347. See Figure A8.2. $\sin 2x = 2\sin x \cos x$. Then $\sin\left(2\cos^{-1}\frac{3}{5}\right) = 2\sin\left(\cos^{-1}\frac{3}{5}\right) \times \cos\left(\cos^{-1}\frac{3}{5}\right) = 2\cdot\frac{4}{5}\cdot\frac{3}{5} = \frac{24}{25}$ $\left(\text{since } \sin\cos^{-1}\frac{3}{5} = \frac{4}{5}\right)$.

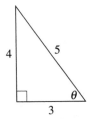

Figure A8.2

348. See Figure A8.3. $\tan 2x = \dfrac{2\tan x}{1-\tan^2 x}$. Then $\tan\left[2\cos^{-1}\left(-\dfrac{4}{5}\right)\right] = \dfrac{2\tan\cos^{-1}\left(-\dfrac{4}{5}\right)}{1-\tan^2\left[\cos^{-1}\left(-\dfrac{4}{5}\right)\right]}$.

Since $\tan\cos^{-1}\left(\dfrac{4}{5}\right) = \dfrac{3}{4}$ and $\tan\cos^{-1}\left(-\dfrac{4}{5}\right) = -\dfrac{3}{4}$ (tan $x < 0$ in quadrant II),

$$\dfrac{2\tan\cos^{-1}\left(-\dfrac{4}{5}\right)}{1-\tan^2\left[\cos^{-1}\left(-\dfrac{4}{5}\right)\right]} = \dfrac{2\left(-\dfrac{3}{4}\right)}{1-\left(-\dfrac{3}{4}\right)^2} = \dfrac{-\dfrac{6}{4}}{1-\dfrac{9}{16}} = -\dfrac{6}{4}\cdot\dfrac{16}{7} = -\dfrac{24}{7}.$$

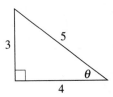

Figure A8.3

349. $\sin\dfrac{x}{2} = \sqrt{\dfrac{1-\cos x}{2}}$ (quadrant I) $= \sqrt{\dfrac{1-\dfrac{1}{3}}{2}} = \sqrt{\dfrac{\dfrac{2}{3}}{2}} = \sqrt{\dfrac{1}{3}}$.

350. $2\sin A\cos B = \sin(A+B) + \sin(A-B)$ $2\cos A\sin B = \sin(A+B) - \sin(A-B)$

$2\cos A\cos B = \cos(A+B) + \cos(A-B)$ $2\sin A\sin B = -\cos(A+B) + \cos(A-B)$

351. $\sin A + \sin B = 2\sin\dfrac{A+B}{2}\cos\dfrac{A-B}{2}$ $\sin A - \sin B = 2\cos\dfrac{A+B}{2}\sin\dfrac{A-B}{2}$

$\cos A + \cos B = 2\cos\dfrac{A+B}{2}\cos\dfrac{A-B}{2}$ $\cos B - \cos A = 2\sin\dfrac{A+B}{2}\sin\dfrac{A-B}{2}$

Be careful with the cos B − cos A identity. Look carefully at A and B.

352. **(A)** $\sin 40°\cos 30° = \frac{1}{2}[\sin(40°+30°)+\sin(40°-30°)] = \frac{1}{2}(\sin 70° + \sin 10°)$.

     (B) $\cos 110°\sin 55° = \frac{1}{2}[\sin(110°+55°)-\sin(110°-55°)] = \frac{1}{2}(\sin 165° - \sin 55°)$.

     (C) $\cos 50°\cos 35° = \frac{1}{2}[\cos(50°+35°)+\cos(50°-35°)] = \frac{1}{2}(\cos 85° + \cos 15°)$.

     (D) $\sin 55°\sin 40° = -\frac{1}{2}[\cos(55°+40°)-\cos(55°-40°)] = -\frac{1}{2}(\cos 95° - \cos 15°)$.

353. $\sin A + \sin B = 2\sin\dfrac{A+B}{2}\cos\dfrac{A-B}{2}$. Letting $A = 20°$ and $B = 15°$, we get $\sin 20° +$

$\sin 15° = 2\sin\dfrac{20°+15°}{2}\cos\dfrac{20°-15°}{2} = 2\sin 17.5°\cos 2.5°$.

354. (A) $\sin 50° + \sin 40° = 2 \sin \frac{1}{2}(50° + 40°) \cos \frac{1}{2}(50° - 40°) = 2 \sin 45° \cos 5°$.

(B) $\sin 70° - \sin 20° = 2 \cos \frac{1}{2}(70° + 20°) \sin \frac{1}{2}(70° - 20°) = 2 \cos 45° \sin 25°$.

(C) $\cos 55° + \cos 25° = 2 \cos \frac{1}{2}(55° + 25°) \cos \frac{1}{2}(55° - 25°) = 2 \cos 40° \cos 15°$.

(D) $\cos 35° - \cos 75° = -2 \sin \frac{1}{2}(35° + 75°) \sin \frac{1}{2}(35° - 75°) = -2 \sin 55° \sin (-20°) = 2 \sin 55° \sin 20°$.

355. $\dfrac{\sin 4A + \sin 2A}{\cos 4A + \cos 2A} = \dfrac{2 \sin \frac{1}{2}(4A + 2A) \cos \frac{1}{2}(4A - 2A)}{2 \cos \frac{1}{2}(4A + 2A) \cos \frac{1}{2}(4A - 2A)} = \dfrac{\sin 3A}{\cos 3A} = \tan 3A$.

356. Letting $x = \dfrac{3\pi}{4}$, we get $1 + \tan\left(\dfrac{3\pi}{4}\right) = 1 + \left[-\tan\left(\dfrac{\pi}{4}\right)\right] = 1 - 1 = 0$. Yes, it is a solution.

357. $\tan\left(\dfrac{\pi}{2}\right)$ is undefined. No, it is not a solution.

358. $4 \cos^2 x = 3$, $\cos^2 x = \frac{3}{4}$, and $\cos x = \pm \dfrac{\sqrt{3}}{\sqrt{4}} = \pm \dfrac{\sqrt{3}}{2}$. If $\cos x = \dfrac{\sqrt{3}}{2}$, then x has a $\dfrac{\pi}{6}$ reference angle. And x must be a quadrant I or IV angle since $\cos x > 0$. $x = \dfrac{\pi}{6}$ or $x = \dfrac{11\pi}{6}$. If $\cos x = -\dfrac{\sqrt{3}}{2}$, x is in quadrant II or III. $x = \dfrac{5\pi}{6}$ or $\dfrac{7\pi}{6}$. *Check*: $4 \cos^2\left(\dfrac{7\pi}{6}\right) \overset{?}{=} 3$, $4\left(-\dfrac{\sqrt{3}}{2}\right)^2 \overset{?}{=} 3$, $3 = 3$. The other three solutions check as well.

359. $2 \sin^2 x = 1$, $\sin^2 x = \frac{1}{2}$, or $\sin x = \pm\sqrt{\dfrac{1}{2}} = \pm 1\sqrt{2} = \pm\dfrac{\sqrt{2}}{2}$. Then $\sin x = \dfrac{\sqrt{2}}{2}$ or $\sin x = -\dfrac{\sqrt{2}}{2}$. If $\sin x = \dfrac{\sqrt{2}}{2}$, x has a $\dfrac{\pi}{4}$ reference angle and is in quadrant I or II. If $\sin x = -\dfrac{\sqrt{2}}{2}$, x has the same reference angle but is in quadrant III or IV. Thus, $x = \dfrac{\pi}{4}, \dfrac{3\pi}{4}, \dfrac{7\pi}{4}, \dfrac{\pi}{4}$. *Check*: $2 \sin^2\left(\dfrac{3\pi}{4}\right) \overset{?}{=} 1$, $2\left(\dfrac{\sqrt{2}}{2}\right)^2 \overset{?}{=} 1$, $1 = 1$. The others check as well.

360. Multiplying the equation by $\sin x$ gives $2 \sin^2 x - 1 = \sin x$, and rearranging, we have $2 \sin^2 x - \sin x - 1 = (2 \sin x + 1)(\sin x - 1) = 0$. From $2 \sin x + 1 = 0$, $\sin x = -\frac{1}{2}$ and $x = \dfrac{7\pi}{6}, \dfrac{11\pi}{6}$; from $\sin x = 1$, $x = \dfrac{\pi}{2}$.

Check. For $x = \dfrac{\pi}{2}$, $2 \sin x - \csc x = 2(1) - 1 = 1$; for $x = \dfrac{7\pi}{6}$ and $\dfrac{11\pi}{6}$, $2 \sin x - \csc x = 2\left(-\dfrac{1}{2}\right) - (-2) = 1$. The solutions are $x = \dfrac{\pi}{2}, \dfrac{7\pi}{6}, \dfrac{11\pi}{6}$.

361. First solution: Putting the equation in the form $\cos x - 1 = \sqrt{3} \sin x$ and squaring, we have $\cos^2 x - 2 \cos x + 1 = 3 \sin^2 x = 3(1 - \cos^2 x)$; then by combining and factoring, $4 \cos^2 x - 2 \cos x - 2 = 2(2 \cos x + 1)(\cos x - 1) = 0$. From $2 \cos x + 1 = 0$, $\cos x = -\frac{1}{2}$ and $x = \dfrac{2\pi}{3}, \dfrac{4\pi}{3}$; from $\cos x - 1 = 0$, $\cos x = 1$ and $x = 0$.

Check: For $x = 0$, $\cos x - \sqrt{3} \sin x = 1 - \sqrt{3}(0) = 1$; for $x = \dfrac{2\pi}{3}$, $\cos x - \sqrt{3} \sin x =$

$-\frac{1}{2} - \sqrt{3}\left(\sqrt{\dfrac{3}{2}}\right) \neq 1$; for $x = \dfrac{4\pi}{3}$, $\cos x - \sqrt{3} \sin x = -\frac{1}{2} - \sqrt{3}\left(-\dfrac{\sqrt{3}}{2}\right) = 1$. The required

solutions are $x = 0, \dfrac{4\pi}{3}$.

Second solution: The left member of the given equation may be put in the form $\sin\theta \cos x + \cos\theta \sin x = \sin(\theta + x)$ in which θ is a known angle, by dividing the given

equation by $r > 0$, $\dfrac{1}{r}\cos x + \left(\dfrac{-\sqrt{3}}{r}\right)\sin x = \dfrac{1}{r}$, and setting $\sin\theta = \dfrac{1}{r}$ and $\cos\theta = \dfrac{-\sqrt{3}}{r}$.

Since $\sin^2\theta + \cos^2\theta = 1$, $\left(\dfrac{1}{r}\right)^2 + \left(-\dfrac{\sqrt{3}}{r}\right)^2 = 1$ and $r = 2$. Now $\sin\theta = \frac{1}{2}$ and $\cos\theta = -\dfrac{\sqrt{3}}{2}$

so that the given equation may be written as $\sin(\theta + x) = \frac{1}{2}$ with $\theta = \dfrac{5\pi}{6}$. Then $\theta + x =$

$\dfrac{5\pi}{6} + x = \arcsin\dfrac{1}{2} = \dfrac{\pi}{6}, \dfrac{5\pi}{6}, \dfrac{13\pi}{6}, \dfrac{17\pi}{6}, \ldots$ and $x = -\dfrac{2\pi}{3}, 0, \dfrac{4\pi}{3}, 2\pi \ldots$. As before, the

required solutions are $x = 0, \dfrac{4\pi}{3}$.

Note that r is the positive square root of the sum of the squares of the coefficients of $\cos x$ and $\sin x$ when the equation is written in the form $a\cos x + b\sin x = 1$, that is,

$$r = \sqrt{a^2 + b^2}$$

The equation will have no solution if $\dfrac{a}{\sqrt{a^2 + b^2}}$ is greater than 1 or less than -1.

362. Substituting $1 + \cos x$ for $2\cos^2\frac{1}{2}x$, the equation becomes $\cos^2 x - \cos x - 1 = 0$;

then $\cos x = \dfrac{(1 \pm \sqrt{5})}{2} \approx 1.6180, -0.6180$. Since $\cos x$ cannot exceed 1, we consider $\cos x = -0.6180$ and obtain the solutions $x = 128°10', 231°50'$.

Note: To solve $\sqrt{2}\cos\frac{1}{2}x = \cos x$ and $\sqrt{2}\cos\frac{1}{2}x = -\cos x$, we square and obtain the equation of this problem. The solution of the first of these equations is $231°50'$, and the solution of the second is $128°10'$.

363. If x is positive, $\alpha = \arccos 2x$ and $\beta = \arcsin x$ terminate in quadrant I; if x is negative, α terminates in quadrant II and β terminates in quadrant IV. Thus, x must be positive.

For x positive, $\sin\beta = x$ and $\cos\beta = \sqrt{1 - x^2}$. Taking the cosine of both members of the given equation, we have

$$\cos(\arccos 2x) = \cos(\arcsin x) = \cos\beta \quad \text{or} \quad 2x = \sqrt{1 - x^2}$$

Squaring we get $4x^2 = 1 - x^2$, $5x^2 = 1$, and $x = \dfrac{\sqrt{5}}{5} \approx 0.4472$.

Check: $\arccos 2x = \arccos 0.8944 = 26°30' = \arcsin 0.4472$, approximating the angle to the nearest 10'.

364. Let $\alpha = \arctan x$ and $\beta = \arctan (1 - x)$; then $\tan \alpha = x$ and $\tan \beta = 1 - x$. Taking the tangent of both members of the given equation gives

$$\tan(\alpha + \beta) = \frac{\tan \alpha + \tan \beta}{1 - \tan \alpha \tan \beta} = \frac{x + (1 - x)}{1 - x(1 - x)} = \frac{1}{1 - x + x^2} = \tan\left(\arctan \frac{4}{3}\right) = \frac{4}{3}$$

Then $3 = 4 - 4x + 4x^2$, $4x^2 - 4x + 1 = (2x - 1)^2 = 0$, $x = \dfrac{1}{2}$.

Check: $\arctan \frac{1}{2} + \arctan (1 - \frac{1}{2}) = 2 \arctan 0.5000 = 53°8'$ and $\arctan \frac{4}{3} = \arctan 1.3333 = 53°8'$.

365. The idea here is to put the equation in a form so that the $\sec^2 \theta = 1 + \tan^2 \theta$ identity can be used. Here is one way: $\sec x = 1 - \tan x$. Squaring both sides of the equation, we get $\sec^2 x = (1 - \tan x)^2 = 1 - 2 \tan x + \tan^2 x$, or $\tan^2 x + 1 = 1 - 2 \tan x + \tan^2 x$. Then $2 \tan x = 0$, $\tan x = 0$, and $x = 0$ or π. It is crucial to check these answers since we squared and may have picked up extraneous solutions. If $x = 0$ then $\sec 0 + \tan 0 = 1 + 0$, and $1 = 1$. If $x = \pi$, then $\sec \pi + \tan \pi = -1 + 0$, and $-1 \neq 1$. Thus, the only solution is $x = 0$.

366. Recall that if $\sin x = r$ ($r \in [-1, 1]$), then $x = 2 k\pi + \sin^{-1} r$ or $x = 2k + (\pi - \sin^{-1} r) = (2k + 1)\pi - \sin^{-1} r$ where $k \in \mathscr{Z}$. Here, using a calculator, we find $\sin^{-1} 0.2977 \approx 0.3023$. Thus, $x = 2k\pi + 0.3023$ or $x = (2k + 1)\pi - 0.3023$ $\forall k \in \mathscr{Z}$.

Chapter 9: Additional Topics in Trigonometry

367. See Figure A9.1. Then $12^2 + a^2 = 13^2$ (by the Pythagorean theorem) and $a^2 = 169 - 144 = 25$, or $a = 5$. Using a calculator, we get $\sin A = $ opposite/hypotenuse $= \frac{5}{13}$, $\angle A = \sin^{-1}\frac{5}{13} \approx 22°37'$; and from a calculator, $\sin B = \frac{12}{13}$, $\angle B = \sin^{-1}\frac{12}{13} \approx 67°23'$.

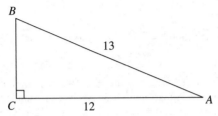

Figure A9.1

368. See Figure A9.2. $\angle A = 90° - 17°50' = 72°10'$. Then $\sin B = \sin 17°50' = \dfrac{b}{3.45}$, or $b = 3.45 \sin 17°50' \approx 1.06$. Thus, $\sin A = \dfrac{a}{3.45}$, $a = 3.45 \sin A \approx 3.28$.

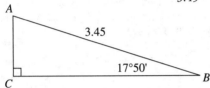

Figure A9.2

369. $a^2 + b^2 = c^2$, $a^2 + 10^2 = (12.6)^2$, $a^2 = 58.76$, or $a \approx 7.67$. Then, $\sin B = \dfrac{10}{12.6}$, so $\angle B = \sin^{-1}\left(\dfrac{10}{12.6}\right) \approx 52°30'$. Thus, $\angle A = 90° - \angle B = 37°30'$. Finally, $\tan B = \dfrac{10}{a}$, $a \tan B = 10$, $a = \dfrac{10}{\tan B} = \dfrac{10}{\tan 52°30'}$, or $a \approx 7.67$ (using a calculator).

370. See Figure A9.3. Then $(\angle AOB) = \dfrac{360°}{6} = 60°$, so $\angle AOD = \dfrac{60°}{2} = 30°$. Since $\sin 30° = \sin(\angle AOD) = \dfrac{AD}{5}$, $AD = 5 \sin 30° = 5 \cdot \frac{1}{2} = 2.5$ m. Thus, $AB = 5$ m, and the perimeter $= P = 6.5 = 30$ m.

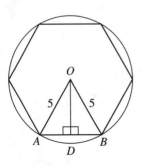

Figure A9.3

371. See Figure A9.4. Let $x =$ number of feet the train climbs. Then $\tan 1°23' = \dfrac{x}{5280}$, and $x = 5280 \tan 1°23' \approx 127.5$ ft.

Figure A9.4

372. See Figure A9.5, where $\angle ACB = 32'$ and $\angle ACD = 16'$. The diameter of the moon $= 2r$. Then $\tan 16' = \dfrac{r}{239,000}$, and $r = (\tan 16') \cdot 239,000$. Thus, the moon's diameter $= 2 \cdot (\tan 16') \cdot 239,000 \approx 2225$ mi.

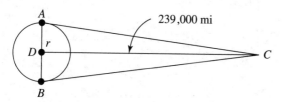

Figure A9.5

373. See Figure A9.6. Then $\tan \theta = \frac{4}{3}$, and $\theta = \arctan \frac{4}{3} \approx 59°$.

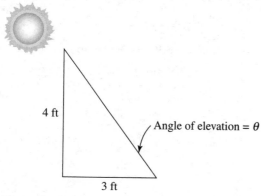

Figure A9.6

374. In any right triangle ABC:

(A) Side $a <$ side c; therefore $\sin A = \dfrac{a}{c} < 1$.

(B) $\sin A = \cos A$ when $\dfrac{a}{c} = \dfrac{b}{c}$; then $a = b$, $A = B$, and $A = 45°$.

(C) $\sin A < 1$ (above) and $\csc A = \dfrac{1}{\sin A} A > 1$.

(D) $\sin A = \dfrac{a}{c}$, $\tan A = \dfrac{a}{b}$, and $b < c$; therefore $\dfrac{a}{c} < \dfrac{a}{b}$ or $\sin A < \tan A$.

(E) $\sin A < \cos A$ when $a < b$; then $A < B$ or $A < 90° - A$, and $A < 45°$.

(F) $\tan A = \dfrac{a}{b} > 1$ when $a > b$; then $A > B$ and $A > 45°$.

375. See Figure A9.7. In any isosceles right triangle ABC, $A = B = 45°$ and $a = b$. Let $a = b = 1$; then $c = \sqrt{1+1} = \sqrt{2}$ and

$$\sin 45° = \frac{1}{\sqrt{2}} = \frac{1}{2}\sqrt{2} \qquad\qquad \cot 45° = 1$$

$$\cos 45° \frac{1}{\sqrt{2}} = \frac{1}{2}\sqrt{2} \qquad\qquad \sec 45° = \sqrt{2}$$

$$\tan 45° = \frac{1}{1} = 1 \qquad\qquad \csc 45° = \sqrt{2}$$

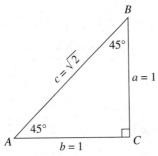

Figure A9.7

376. See Figure A9.8. In any equilateral triangle ABD, each angle is 60°. The bisector of any angle, as B, is the perpendicular bisector of the opposite side. Let the sides of the equilateral triangle be 2 units long. Then in the right triangle ABC, $AB = 2$, $AC = 1$, and $BC = \sqrt{2^2 - 1^2} = \sqrt{3}$.

$$\sin 30° = \frac{1}{2} = \cos 60° \qquad\qquad \cot 30° = \sqrt{3} = \tan 60°$$

$$\cos 30° = \frac{\sqrt{3}}{2} = \sin 60° \qquad\qquad \sec 30° = \frac{2}{\sqrt{3}} = \frac{2\sqrt{3}}{3} = \csc 60°$$

$$\tan 30° = \frac{1}{\sqrt{3}} = \frac{\sqrt{3}}{3} = \cot 60° \qquad\qquad \csc 30° = 2 = \sec 60°$$

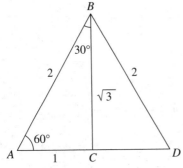

Figure A9.8

377. In Figure A9.9, $A = 20°$ and $CB = 150$. Then $\cot A = \dfrac{AC}{CB}$ and $AC = CB \cot A$ = 150 cot 20° =150(2.7) = 405 ft.

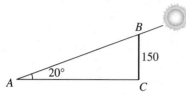

Figure A9.9

378. In Figure A9.10, $CB = 100$ and $AC = 120$. Then $\tan A = \dfrac{CB}{AC} = \dfrac{100}{120} = 0.83$ and $A = 40°$.

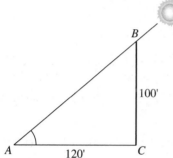

Figure A9.10

379. In Figure A9.11, OC bisects $\angle AOB$. Then $BC = AC$ and OAC is a right triangle. In $\triangle OAC$, $\sin \angle COA = \dfrac{AC}{OA}$ and $AC = OA \sin \angle COA = 20 \sin 75° = 20(0.97) = 19.4$. Then $BA = 38.8$, and the length of the chord is 39 m.

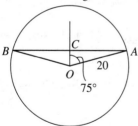

Figure A9.11

380. In right triangle ABC, $\cot A = \dfrac{AC}{CB}$; then $AC = CB \cot A$ or $DC + 75 = CB \cot 20°$. In right triangle DBC, $\cot D = \dfrac{DC}{CB}$; then $DC = CB \cot 40°$. Then $DC = CB \cot 20° - 75 = CB \cot 40°$, $CB(\cot 20° - \cot 40°) = 75$, $CB(2.7 - 1.2) = 75$, and $CB = \dfrac{75}{1.5} = 50$ ft.

381. $\alpha = 180° - 12°40' - 100° = 67°20'$. The law of sines is $\dfrac{\sin \alpha}{a} = \dfrac{\sin \beta}{b}$. Then $a = \dfrac{b \sin \alpha}{\sin \beta} = \dfrac{13.1 \sin 67°20'}{\sin 12°40'} = 55.1$.

382. $\alpha = 180° - 54°30' - 27°30' = 98°$. Then from $\dfrac{\sin \beta}{b} = \dfrac{\sin \alpha}{a}$, $b = \dfrac{a \sin \beta}{\sin \alpha} = \dfrac{9.27 \sin 27°30'}{\sin 98°} = 4.32$. $\dfrac{\sin \gamma}{c} = \dfrac{\sin \alpha}{a}$, so $c = \dfrac{a \sin \gamma}{\sin \alpha} = \dfrac{9.27 \sin 54°30'}{\sin 98°} = 7.62$.

383. Here α is acute. Then from $\sin \dfrac{\alpha}{a} = \sin \dfrac{\gamma}{c}$, $\sin \alpha = a \sin \dfrac{\gamma}{c} = 50 \sin \dfrac{30°}{40} = 0.6250$. Thus, $\alpha = \mathrm{Sin}^{-1} 0.6250 = 39°$.

$\beta = 180° - 39° - 30° = 111°$. Then from $\dfrac{\sin \beta}{b} = \dfrac{\sin \gamma}{c}$, $b = \dfrac{40 \sin 111°}{\sin 30°} = 75$.

384. Then from $\dfrac{\sin \alpha}{a} = \dfrac{\sin \beta}{b}$, $\sin \beta = \dfrac{23 \sin 41°}{14} = 1.078$. Thus $\beta = \sin^{-1} 1.078$; there is no solution. Draw the triangle; do you see why there is no solution?

385. From the law of cosines $a^2 = b^2 + c^2 - 2bc \cos \alpha$, we have $a^2 = (7.03)^2 + (7)^2 - 2(7.03)(7) \cos 50°40' = 36.03925 \cdots$, or $a = 6.00$.

From the law of sines $\dfrac{a}{\sin \alpha} = \dfrac{b}{\sin \beta}$, $\sin \beta = 7.03 \sin \dfrac{50°40'}{6} = 0.9063$. Then $\beta = \mathrm{Sin}^{-1} 0.9063 = 65°0'$. (Do *not* choose $180° - \mathrm{Sin}^{-1} 0.9063$ since there cannot be two obtuse angles in a triangle.)

386. $c^2 = a^2 + b^2 - 2ab \cos \gamma = (5.73)^2 + (10.2)^2 - 2(5.73)(10.2) \cos (120°20') = 195.90686 \cdots$. Thus, $c = 14.0$.

From $\dfrac{\sin \beta}{b} = \sin \dfrac{\gamma}{c}$, $\sin \beta = b \sin \dfrac{\gamma}{c}$. Then $\beta = \mathrm{Sin}^{-1} 0.6288 = 39°0'$. (Do *not* use $180° - \mathrm{Sin}^{-1} 0.6288$ since γ in obtuse.)

387. $a^2 = b^2 + c^2 - 2bc \cos \alpha$. Then $\cos \alpha = \dfrac{[(10.0)^2 + (9.00)^2 - (4.00)^2]}{180} = 0.9167$. Thus, $\alpha = \cos^{-1} 0.9167 = 23°30'$. (Note that, given the three lengths we have, α and γ are both less than $90°$.)

From $\sin \dfrac{\gamma}{c} = \sin \dfrac{\alpha}{a}$, $\sin \gamma = 9.00 \sin \dfrac{23°30'}{4.00} = 0.8972$. Thus, $\gamma = \sin^{-1} 0.8972$ ($\gamma < 90°$) = $63°50'$.

388. See Figure A9.12. The "length" of the vector is 3 (magnitude), and we move $40°$ in the easterly direction from north (y axis).

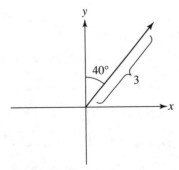

Figure A9.12

389. See Figure A9.13. $AB = 5$, and we move 60° west from the "south axis."

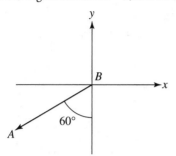

Figure A9.13

390. The magnitude of $v = (4, 5)$, written $|v|$ or $|(4, 5)| = \sqrt{4^2 + 5^2} = \sqrt{16 + 25} = \sqrt{41}$.

391. $|(6, -14)| = \sqrt{6^2 + (-14)^2} = \sqrt{36 + 196} = \sqrt{232} = 2\sqrt{58}$.

392. If $v = (x, y)$, then θ is the direction of v, where $\tan \theta = \dfrac{y}{x}$. Here $\tan \theta = -\dfrac{3}{4}$ and $\theta = 323°8'$ (use a calculator).

393. See question 392. $\tan \theta = \dfrac{-1}{-2} = \dfrac{1}{2}$. Then $\theta = 206°34'$.

394. **(A)** Recall that $k(x, y) = (kx, ky)$. Thus $2a = 2(1, 0) = (2, 0)$.

   (B) $\dfrac{2}{3}b = \dfrac{2}{3}(3, 0) = \left(\dfrac{2}{3} \cdot 3 \cdot \dfrac{2}{3} \cdot 0\right) = (2, 0) = 2a$ (see question 394A).

   (C) Recall that $(x_1, y_1) + (x_2, y_2) = (x_1 + x_2, y_1 + y_2)$. Thus, $a + c = (1, 0) + (4, 6) = (1 + 4, 0 + 6) = (5, 6)$.

   (D) Recall that $a - b = a + (-b)$. Here, $d = (4, 9)$, so $-d = -1d = (-4, -9)$ and $a - d = a + (-d) = (1, 0) + (-4, -9) = (1 - 4, 0 - 9) = (-3, -9)$.

   (E) If $x = (x_1, y_1)$ and $y = (x_2, y_2)$, then $x \cdot y = x_1 x_2 + y_1 y_2 = $ the dot product. Here, $a \cdot c = (1, 0) \cdot (4, 6) = 1 \cdot 4 + 0 \cdot 6 = 4 + 0 = 4$.

395. $i = (1, 0)$ and $j = (0, 1)$. Then $(4, 7) = 4(1, 0) + 7(0, 1) = 4i + 7j$.

396. $|(6, 4)| = \sqrt{36 + 16} = \sqrt{52}$. Then $|A| = 1$, where $A = \left(\dfrac{6}{\sqrt{52}}, \dfrac{4}{\sqrt{52}}\right)$, and the direction of $A = $ the direction of $(6, 4)$.

397. $(1, 0) \cdot (\sqrt{2}, \sqrt{2}) = 1 \cdot \sqrt{2} + 0 \cdot \sqrt{2} = \sqrt{2}$. Also $(1, 0) \cdot (\sqrt{2}, \sqrt{2}) = |(1, 0)||(\sqrt{2}, \sqrt{2})| \times \cos \theta$. Thus $\sqrt{2} = 1 \cdot 2 \cos \theta$, $\cos \theta = \dfrac{\sqrt{2}}{2}$, and $\theta = \dfrac{\pi}{4}$.

398. See Figure A9.14. We need the component of ray XY parallel to the driveway; i.e., we need $|TX|$. Since $\dfrac{|TX|}{2000} = \sin 5°$, $|TX| = 2000 \sin 5° = 174.5$ lb. See question 401.

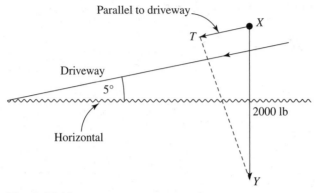

Figure A9.14

399. The speed of the airplane is 240 mi/h = 240(5280)/[60(60)] ft/s = 352 ft/s. In Figure A9.15, vector, AB represents the velocity of the airplane, vector AC represents the initial velocity of the bullet, and vector AD represents the resultant velocity of the bullet. In right triangle ACD, $AD \sqrt{(352)^2 + (2750)^2} = 2770$ ft/s, $\tan \angle CAD = \frac{352}{2750} = 0.1280$, and $\angle CAD = 7°20'$. Thus, the bullet travels at 2770 ft/s along a path making an angle of 82°40' with the path of the airplane.

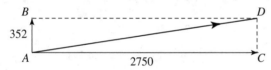

Figure A9.15

400. (A) Refer to Figure 9.3 in the question section. In right triangle OAB, $OB = \sqrt{(475)^2 + (125)^2} = 491$, $\tan \theta = \frac{125}{475} = 0.2632$, and $\theta = 14°40'$. Thus the boat moves at 491 ft/min in the direction S75°20′E.

  (B) Refer to Figure A9.16. In right triangle OAB, $\sin \theta = \frac{125}{475} = 0.2632$ and $\theta = 15°20'$. Thus the boat must be headed N74°40′E, and its speed in that direction is $OB = \sqrt{(475)^2 - (125)^2} = 458$ ft/min.

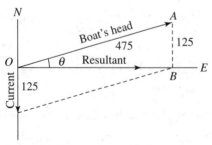

Figure A9.16

401. Refer to Figure A9.17. Resolve the weight **W** of the block into components $\mathbf{F_1}$ and $\mathbf{F_2}$, respectively, parallel and perpendicular to the ramp. $\mathbf{F_1}$ is the force tending to move the block down the ramp, and $\mathbf{F_2}$ is the force of the block on the ramp.

$$\mathbf{F}_1 = \mathbf{W} \sin 29° = 500(0.4848) = 242 \text{ lb}$$

$$\mathbf{F}_2 = \mathbf{W} \cos 29° = 500(0.8746) = 437 \text{ lb}$$

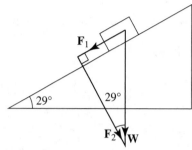

Figure A9.17

Chapter 10: Conic Sections

402. If $(x-h)^2 + (y-k)^2 = r^2$, then (h, k) is the center and r is the radius. Here, $C(0, 0)$, $r = \sqrt{100} = 10$.

403. $7(x^2 + 2x + 1) + 7(y^2 - 8y + 16) = 25 + 7 + 112 = 144$, $(x+1)^2 + (y-4)^2 = \frac{144}{7}$, and $C(-1, 4)$, $r = \frac{12}{\sqrt{7}}$.

404. $x^2 - 6x + 9 + y^2 + 8y + 16 = 11 + 9 + 16$, $(x-3)^2 + (y+4)^2 = 36$, and $C(3, -4)$, $r = 6$.

405. $(x+2)^2 + (y+3)^2 = (\sqrt{7})^2$, or $(x+2)^2 + (y+3)^2 = 7$.

406. $(x-h)^2 + (y-k)^2 = r^2$. Using $(0, 0)$: $(-h)^2 + (-k)^2 = r^2$, and $h^2 + k^2 = r^2$. Using $(1, 1)$: $(1-h)^2 + (1-k)^2 = r^2$, $1 - 2h + h^2 + 1 - 2k + k^2 = r^2$, and $2 + h^2 - 2h - 2k + k^2 = r^2$. Using $(1, 2)$: $(1-h)^2 + (2-k)^2 = r^2$, $1 - 2h + h^2 + 4 - 4k + k^2 = r^2$, and $5 + h^2 - 2h - 4k + k^2 = r^2$. From these equations we have $-2 + 2h + 2k = 0$ and $-3 + 2k = 0$, so $k = \frac{3}{2}$.

Thus, $-2 + 2h + 3 = 0$, $2h = 1$, or $h = \frac{1}{2}$. Then, $\frac{1}{4} + \frac{1}{4} = r^2$, $r^2 = \frac{1}{2}$, $r = \dfrac{1}{\sqrt{2}} = \dfrac{\sqrt{2}}{2}$. Thus, $C = \left(\dfrac{1}{2}, \dfrac{1}{2}\right)$, $r = \dfrac{\sqrt{2}}{2}$.

407. $y^2 = 40x$ means $y^2 = 4ax$ where $a = 10$. Focus: $(10, 0)$. Directrix: $x = -10$.

408. $4a = -16$, or $a = -4$. Focus: $(0, -4)$. Directrix: $y = 4$.

409. See Figure A10.1. $8 = 4a$, or $a = 2$. Focus: $(2, 0)$. Directrix: $x = -2$.

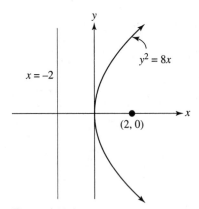

Figure A10.1

410. See Figure A10.2. $10 = 4a$, or $a = \frac{5}{2}$ Focus: $\left(0, \frac{5}{2}\right)$. Directrix: $y = -\frac{5}{2}$.

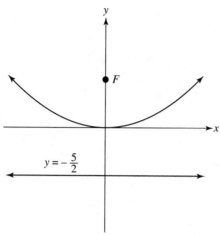

Figure A10.2

411. $(x^2 - 6x + 9) + 8y = -25 + 9$, $(x - 3)^2 = -8y - 16$, $(x - 3)^2 = -8\,(y + 2)$, so $h = 3$, $k = -2$. Vertex: $(3, -2)$.

412. $y^2 + 2y + 1 = 16x - 49 + 1 = 16x - 48$, $(y + 1)^2 = 16(x - 3)$, so $k = -1$, $h = 3$. Vertex: $(3, -1)$.

413. This is of the form $\dfrac{x^2}{a^2} + \dfrac{y^2}{b^2} = 1$, where $a > b > 0$. Thus, the foci are at $(\pm c, 0)$, where $c^2 = a^2 - b^2$. Since $c^2 = 25 - 4 = 21$, $c = \pm\sqrt{21}$, and the foci are at $(\pm\sqrt{21}, 0)$. The length of the major axis $= 2a = 10$ $(a = 5)$; the length of the minor axis $= 2b = 4$ $(b = 2)$.

414. Here, $\dfrac{x^2}{b^2} + \dfrac{y^2}{a^2} = 1$, $a > b > 0$. Then the foci are at $(0, \pm c)$, where $c^2 = a^2 - b^2$. The major axis length $= 2a$; the minor axis length $= 2b$. Since $c^2 = a^2 - b^2 = 25 - 4 = 21$, the foci are at $(0, \pm\sqrt{21})$. The length of the major axis $= 10$ $(a = 5)$; the length of the minor axis $= 4$ $(b = 2)$.

415. See Figure A10.3 and question 413. $a^2 = 25$ and $b^2 = 4$, where $\dfrac{x^2}{a^2} + \dfrac{y^2}{b^2} = 1$. Thus, the x intercepts $= \pm\,a$ and the y intercepts $= \pm\,b$. F_1 and F_2 are the foci.

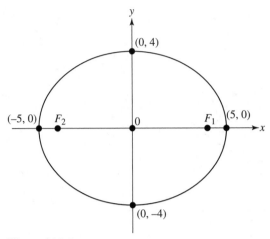

Figure A10.3

416. See Figure A10.4 and question 414. $a^2 = 25$ and $b^2 = 4$. Then the x intercepts $= \pm 2$, and the y intercepts $= \pm 5$.

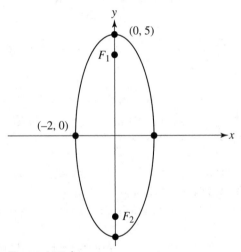

Figure A10.4

417. $(x^2 - 6x + 9) + 4(y^2 + 8y + 16) = -69 + 9 + 64$, $(x - 3)^2 + 4(y + 4)^2 = 4$, and $\dfrac{(x - 3)^2}{4} + (y + 4)^2 = 1$. Center: $(3, -4)$.

418. See Figure A10.5 and question 417. $\dfrac{(x - 3)^2}{4} + \dfrac{(y + 4)^2}{1} = 1$. Then $a^2 = 4$, $a = 2$; and $b^2 = 1$, $b = 1$. Thus the vertices must be at a distance of $a\ (= 2)$ from the center. The vertices are $(5, -4)$ and $(1, -4)$. The foci are $\sqrt{3}$ from the center. Foci: $(3 \pm \sqrt{3}, -4)$

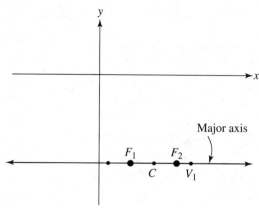

Figure A10.5

419. If $\dfrac{x^2}{a^2} - \dfrac{y^2}{b^2} = 1$, then the foci are $(\pm c, 0)$, where $c^2 = a^2 + b^2$, and the lengths of the transverse and conjugate axes are $2a$ and $2b$, respectively. Here $a^2 = 9$, $b^2 = 4$, and $c^2 = 9 + 4 = 13$. Thus, the foci are $(\pm\sqrt{13}, 0)$. The length of the transverse axis $= 6 \,(= 2a)$, and the length of the conjugate axis $= 4 \,(= 2b)$.

420. Here, $\dfrac{y^2}{a^2} - \dfrac{x^2}{b^2} = 1$. If $c^2 = a^2 + b^2$, then the foci are at $(0, \pm c)$; the transverse axis length is $2a$, the conjugate axis length is $2b$. Then $c^2 = 4 + 9 = 13$. The foci are at $(0, \pm\sqrt{13})$; the length of the transverse axis $= 4$; and the length of the conjugate axis $= 6$.

421. Divide by 16: $\dfrac{x^2}{4} - \dfrac{y^2}{16} = 1$. Then $c^2 = a^2 + b^2 = 20$, and $c = \pm\sqrt{20} = \pm 2\sqrt{5}$. The foci are at $(\pm 2\sqrt{5}, 0)$; the length of the transverse axis $= 4$; and the length of the conjugate axis $= 8$.

422. $\dfrac{y^2}{8} - \dfrac{x^2}{12} = 1$. Then $c^2 = 8 + 12 = 20$, $c = 2\sqrt{5}$. The foci are at $(0, \pm 2\sqrt{5})$; the length of the transverse axis $= 2(2\sqrt{2}) = 4\sqrt{2}$; and the length of the conjugate axis $= 2(2\sqrt{3}) = 4\sqrt{3}$.

423. See Figure A10.6 and question 419. Vertices are V_1, V_2 $(\pm 3, 0)$ (when $y = 0$). The center is at $C\,(0, 0)$; the length of the transverse axis $= 6$; and the length of the conjugate axis $= 4$.

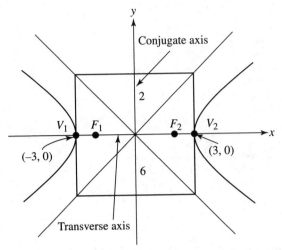

Figure A10.6

424. See Figure A10.7 and question 422. Then $\dfrac{y^2}{8} - \dfrac{x^2}{12} = 1$.

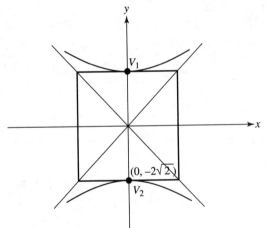

Figure A10.7

425. The asymptotes for an equation of the form $\dfrac{x^2}{a^2} - \dfrac{y^2}{b^2} = 1$ are $y = \pm\left(\dfrac{b}{a}\right)x$; here $y = \pm\dfrac{2}{1}x$ or $y = 2x$ and $y = -2x$.

426. This hyperbola is of the form $\dfrac{y^2}{a^2} - \dfrac{x^2}{b^2} = 1$. Here, $y = \pm\left(\dfrac{\sqrt{3}}{\sqrt{2}}\right)x$, or $y = \pm\sqrt{\dfrac{3}{2}}x$.

427. $(x^2 + 6x + 9) - 4(y^2 - 4y + 4) = 11 + 9 - 16$, $(x + 3)^2 - 4(y - 2)^2 = 4$, and $\dfrac{(x+3)^2}{4} - \dfrac{(y-2)^2}{1} = 1$. The center is at $(-3, 2)$ since $h = -3$, $k = 2$.

428. $144(x^2 - 4x + 4) - 25(y^2 - 8y + 16) = -3776 + 576 - 400$, $144(x - 2)^2 - 25(y - 4)^2 = -3600$, $\dfrac{25(y-4)^2}{3600} - \dfrac{144(x-2)^2}{3600} = 1$, $\dfrac{(y-4)^2}{\frac{3600}{25}} - \dfrac{(x-2)^2}{\frac{3600}{144}} = 1$, $\dfrac{(y-4)^2}{144} - \dfrac{(x-2)^2}{25} = 1$. The center is at (2, 4).

429. See question 427. Then $\dfrac{(x+3)^2}{4} - \dfrac{(y-2)^2}{1} = 1$. Thus, $a = 2$, $b = 1$, and $c = \sqrt{a^2 + b^2} = \sqrt{5}$. The vertices are on the transverse axis at a distance a from the center $(-3, 2)$. Thus, the vertices are $(-1, 2)$ and $(-5, 2)$. The foci are at a distance c from the center $(-3, 2)$. Thus, the foci are $(-3 \pm \sqrt{5}, 2)$.

430. See question 428. Then $\dfrac{(y-4)^2}{144} - \dfrac{(x-2)^2}{25} = 1$. Thus, $a = 12$, $b = 5$, and $c = \sqrt{a^2 + b^2} = \sqrt{144 + 25} = 13$. Then the vertices are $V_1 = (2, 4 + 12) = (2, 16)$ and $V_2 = (2, 4 - 12) = (2, -8)$. The foci are $F_1 = (2, 4 + 13) = (2, 17)$ and $F_2 = (2, 4 - 13) = (2, -9)$.

431. From the standard form $(x - 5)^2 + (y + 4)^2 = 36$, the locus is a circle with center at $C(5, -4)$ and radius 6.

432. Here we have $(x + 2)^2 + (y - 3)^2 = -11$; the locus is imaginary.

433. The parabola opens to the right $(p > 0)$ with vertex at $V(0, 0)$. The equation of its axis is $y = 0$. Moving from V to the right along the axis a distance $|p| = 4$, we locate the focus at $F(4, 0)$. Moving from V to the left along the axis a distance $|p| = 4$, we locate the point $D(-4, 0)$. Since the directrix passes through D perpendicular to the axis, its equation is $x + 4 = 0$. See Figure A10.8.

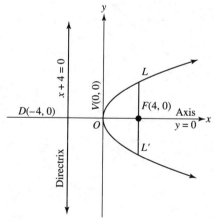

Figure A10.8

434. The parabola opens downward $(p < 0)$ with vertex at $V(0, 0)$. The equation of its axis is $x = 0$. Moving from V downward along the axis a distance $|p| = \frac{9}{4}$, we locate the focus at $F\left(0, -\dfrac{9}{4}\right)$. Moving from V upward along the axis a distance $|p| = \dfrac{9}{4}$, we locate the point $D\left(0, \dfrac{9}{4}\right)$; the equation of the directrix is $4y - 9 = 0$. See Figure A10.9.

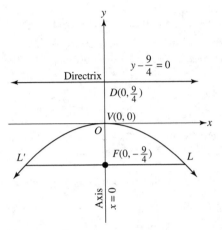

Figure A10.9

435. Here $(x - 1)^2 = 12(y - 2)$. The parabola opens upward ($p > 0$) with vertex at $V(1, 2)$. The equation of its axis is $x - 1 = 0$. Since $|p| = 3$, the focus is at $F(1, 5)$, and the equation of the directrix is $y + 1 = 0$. See Figure A10.10.

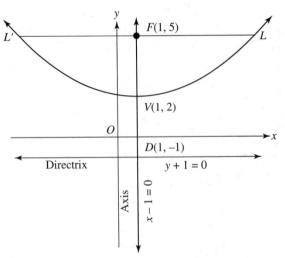

Figure A10.10

436. The equation of the conjugate hyperbola is $16y^2 - 25x^2 = 400$. The common asymptotes have equations $y = \pm\dfrac{5x}{4}$. The vertices of $25x^2 - 16y^2 = 400$ are at $(\pm4, 0)$. The vertices of $16y^2 - 25x^2 = 400$ are at $(0, \pm5)$. The curves are shown in Figure A10.11.

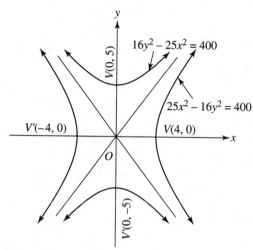

Figure A10.11

Chapter 11: The Complex Numbers

437. **(A)** Remember that, given $a + bi$ where $a, b \in \mathfrak{R}$, $a + bi$ corresponds to the point (a, b) in the Cartesian plane. Thus, $2 + i$ corresponds to $(2, 1)$. See Figure A11.1.

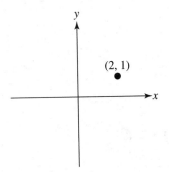

Figure A11.1

(B) See question 437A. $2 + 3i$ corresponds to the point $(2, 3)$. See Figure A11.2.

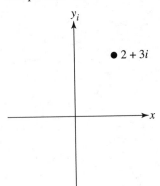

Figure A11.2

(C) $2 - i$ corresponds to $(2, -1)$. See Figure A11.3.

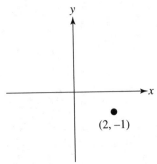

Figure A11.3

(D) Since $i = 0 + 1i$, i corresponds to (0, 1). See Figure A11.4.

Figure A11.4

(E) $6 = 6 + 0i$ and corresponds to the point (6, 0). See Figure A11.5.

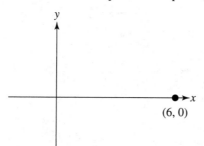

Figure A11.5

438. $d(A, P) = 4$ and $\theta = 0$ (θ is the angle between AP and the positive x axis), so $4 + 0i = 4(\cos 0 + i \sin 0)$. See Figure A11.6.

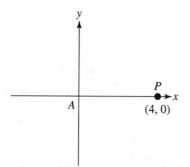

Figure A11.6

439. $d(A,P) = \sqrt{(2\sqrt{3})^2 + 4} = \sqrt{12 + 4} = 4 = r$. $\tan\theta = \dfrac{b}{a} = \dfrac{-2}{2\sqrt{3}} = \dfrac{-1}{\sqrt{3}} = \dfrac{-\sqrt{3}}{3}$, so $\theta = \dfrac{11\pi}{6}$. $2\sqrt{3} - 2i = 4\left[\cos\left(\dfrac{11\pi}{6}\right) + i \sin\left(\dfrac{11\pi}{6}\right)\right]$. See Figure A11.7.

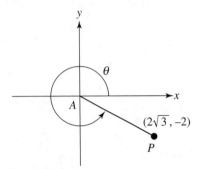

Figure A11.7

440. $r = \sqrt{5^2 + 5^2} = \sqrt{50} = 5\sqrt{2}$. $\tan\theta = \dfrac{5}{5} = 1$, so $\theta = \dfrac{\pi}{4}$. $5 + 5i = 5\sqrt{2}\left[\cos\left(\dfrac{\pi}{4}\right) + i\sin\left(\dfrac{\pi}{4}\right)\right]$. See Figure A11.8.

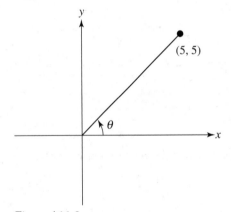

Figure A11.8

441. Recall that the product is found by multiplying moduli and adding corresponding arguments. In this case, $\sqrt{2}\cdot 3 = 3\sqrt{2}$, and $\dfrac{\pi}{4} + \dfrac{\pi}{2} = \dfrac{3\pi}{4}$. So the answer is $3\sqrt{2}\left[\cos\left(\dfrac{3\pi}{4}\right) + i\sin\left(\dfrac{3\pi}{4}\right)\right]$.

442. $\sqrt{2}\cos\left(\dfrac{5\pi}{4}\right) = \sqrt{2}\cdot\left(-\dfrac{\sqrt{2}}{2}\right) = -1$, and $\sqrt{2}\left[i\sin\left(\dfrac{5\pi}{4}\right)\right] = \sqrt{2}\cdot i\cdot-\dfrac{\sqrt{2}}{2} = -i$. So the answer is $-1 - i$.

443. $4\cos 0 = 4\cdot 1 = 4$, and $i\cdot 4\sin 0 = i\cdot 0 = 0$. So the answer is $4 + 0i = 4$. See question 438.

444. $4\cos\left(\dfrac{11\pi}{6}\right) = 4\cdot\left(\dfrac{\sqrt{3}}{2}\right) = 2\sqrt{3}$, and $i\cdot 4\sin\left(\dfrac{11\pi}{6}\right) = -2i$. So the answer is $2\sqrt{3} - 2i$.
See question 439.

445. The product is $6\left[\cos\left(\dfrac{4\pi}{6}\right)+i\sin\left(\dfrac{4\pi}{6}\right)\right]=6\left(-\dfrac{1}{2}\right)+6\left(i\cdot\dfrac{\sqrt{3}}{2}\right)=-3+3i\sqrt{3}$.

446. The quotient is $4[\cos(-150°)+i\sin(-150°)]=-2\sqrt{3}-2i$.

447. $1+i=\sqrt{2}\left[\cos\left(\dfrac{\pi}{4}\right)+i\sin\left(\dfrac{\pi}{4}\right)\right]$, and $\sqrt{2}-i\sqrt{2}=2\left[\cos\left(\dfrac{7\pi}{4}\right)+i\sin\left(\dfrac{7\pi}{4}\right)\right]$. The product is $2\sqrt{2}(\cos 2\pi+i\sin 2\pi)=2\sqrt{2}\cdot 1=2\sqrt{2}$.

448. Let $n\in\mathscr{L}$. Then $[r(\cos\theta+i\sin\theta)]^n=r^n(\cos n\theta+i\sin n\theta)$.

449. Since $3^3=27$, and $10°\cdot 3=30°$, the answer is $27(\cos 30°+i\sin 30°)$.

450. $\left[\cos\left(\dfrac{\pi}{3}\right)+i\sin\left(\dfrac{\pi}{3}\right)\right]^5=\cos\left(\dfrac{5\pi}{3}\right)+i\sin\left(\dfrac{5\pi}{3}\right)=\dfrac{1}{2}-\dfrac{\sqrt{3}}{2}i$.

451. $\tan\theta=\dfrac{b}{a}=\dfrac{1}{\sqrt{3}}\cdot\sqrt{3}+i=2\left[\cos\left(\dfrac{\pi}{6}\right)+i\sin\left(\dfrac{\pi}{6}\right)\right]$. Then $(\sqrt{3}+i)^{12}=(2)^{12}(\cos 2\pi+i\sin 2\pi)=4096+0i$. See Figure A11.9.

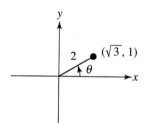

Figure A11.9

452. See Figure A11.10. Recall that if $z\ (\neq 0)=r(\cos\theta+i\sin\theta)$, then z has n nth roots given by the formula

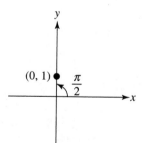

Figure A11.10

$$r^{\frac{1}{n}}\left(\cos\dfrac{\theta+2\pi k}{n}+i\sin\dfrac{\theta+2\pi k}{n}\right)\qquad\text{for}\qquad k=0,1,\ldots,n-1$$

Since $i = 0 + i = 1\left[\cos\left(\dfrac{\pi}{2}\right) + i\sin\left(\dfrac{\pi}{2}\right)\right]$, the square roots of i are $1^{\frac{1}{2}}\left(\cos\dfrac{\theta + 2\pi k}{n} + i\sin\dfrac{\theta + 2\pi k}{n}\right)$

for $k = 0$, 1 where $0 = \left(\dfrac{\pi}{2}\right)$, $n = 2$. The two roots are

$$w_1 = \cos\frac{\frac{\pi}{2}}{2} + i\sin\frac{\frac{\pi}{2}}{2} = \cos\frac{\pi}{4} + i\sin\frac{\pi}{4} = \frac{\sqrt{2}}{2} + i\frac{\sqrt{2}}{2}$$

and

$$w_2 = \cos\frac{\frac{\pi}{2} + 2\pi}{2} + i\sin\frac{\frac{\pi}{2} + 2\pi}{2} = \frac{-\sqrt{2}}{2} - i\frac{\sqrt{2}}{2}$$

453. See Figure A11.11. $-8 = -8 + 0i = 8(\cos\pi + i\sin\pi)$. The cube roots are

$$w_1 = 8^{\frac{1}{3}}\left(\cos\frac{\pi}{3} + i\sin\frac{\pi}{3}\right) = 1 + \sqrt{3}i, \quad w_2 = 8^{\frac{1}{3}}\left(\cos\frac{2\pi + \pi}{3} + i\sin\frac{2\pi + \pi}{3}\right) = -2, \text{ and}$$

$$w_3 = 8^{\frac{1}{3}}\left(\cos\frac{4\pi + \pi}{3} + i\sin\frac{4\pi + \pi}{3}\right) = 1 - \sqrt{3}i.$$

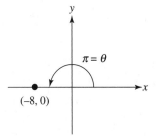

Figure A11.11

454. See Figure A11.12. $1 + i\sqrt{3} = 2\left(\cos\dfrac{\pi}{3} + i\sin\dfrac{\pi}{3}\right)$. Then $w_1 = 2^{\frac{1}{2}}\left(\cos\dfrac{\frac{\pi}{3}}{2} + i\sin\dfrac{\frac{\pi}{3}}{2}\right) =$

$\dfrac{\sqrt{6}}{2} + \dfrac{i\sqrt{2}}{2}$ and $w_2 = 2^{\frac{1}{2}}\left(\cos\dfrac{2\pi + \frac{\pi}{3}}{2} + i\sin\dfrac{2\pi + \frac{\pi}{3}}{2}\right) = \dfrac{\sqrt{6}}{2} - \dfrac{i\sqrt{2}}{2}$.

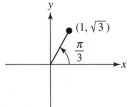

Figure A11.12

455. $x^3 + 1 = 0$; then $x^3 = -1$, and x can be any one of the three cube roots of -1. $-1 = -1 + 0i = 1(\cos\pi + i\sin\pi)$. Then $w_1 = -1$, $w_2 = \dfrac{1 + \sqrt{3}i}{2}$, and $w_3 = \dfrac{1 - \sqrt{3}i}{2}$.

456. If $x^2 - (1 + i\sqrt{3}) = 0$, $x =$ square roots of $1 + i\sqrt{3}$. Then $w_1 = \dfrac{\sqrt{6}}{2} + i\dfrac{\sqrt{2}}{2}$, and $w_2 = -\dfrac{\sqrt{6}}{2} - i\dfrac{\sqrt{2}}{2}$. (See question 454.)

Chapter 12: Sequences, Series, and Probability

457. Let $n = 1$, 2, 3, and 4 to generate the first four terms: $1 - 3$, $2 - 3$, $3 - 3$, $4 - 3$; or $-2, -1, 0, 1$.

458. $\dfrac{1}{4^{1-2}+1}, \dfrac{1}{4^{2-2}+1}, \dfrac{1}{4^{3-2}+1}, \dfrac{1}{4^{4-2}+1}$; or $\dfrac{4}{5}, 1, \dfrac{1}{5}, \dfrac{1}{17}$.

459. When $n = 1$, $(-1)^n = -1^1 = -1$. Then the sign of the $n + 1$ term is -1, and the signs alternate $-1, +1, -1$, etc. Thus, the sequence is $-\dfrac{1}{3}, \dfrac{1}{4}, -\dfrac{1}{5}, \dfrac{1}{6}$.

460. Notice that each term's numerator is 1 less than its denominator. Also the numerator of the first term is 1. We try $\dfrac{n}{n+1}$. Checking, we see that this is not the general term! Check it. However, when we notice that the numerators are all odd and the denominators even, we find the general term: $\dfrac{2n-1}{2n}$. Check this; it is correct.

461. This is the famous *Fibonacci sequence*. $A_1 = 1$; $A_2 = 1$; $A_3 = A_2 + A_1 = 1 + 1 = 2$; $A_4 = A_3 + A_2 = 2 + 1 = 3$. The terms are 1, 1, 2, 3.

462. How can we get 5 from 1 and 1? How about $3(1) + 2(1)$? Note that $17 = 3(5) + 2(1)$, etc. $A_1 = 1$, $A_2 = 1$, $A_{n+2} = 3A_{n+1} + 2A_n$.

463. When $i = 1$, $(-1)^2 (x^1 - 1) = x - 1$. When $i = 2$, $(-1)^3 (x^2 - 1) = -(x^2 - 1) = 1 - x^2$. When $i = 3$, $(-1)^4 (x^3 - 1) = x^3 - 1$. Thus, we have $(x - 1) + (1 - x^2) + (x^3 - 1)$.

464. These are the consecutive odd numbers $2i - 1$ beginning with $i = 1$, alternating with a positive lead term: $\sum\limits_{i=1}^{6}(-1)^{i+1}(2i-1)$.

465. Geometric; ratio $= \dfrac{1}{2}$.

466. Arithmetic; difference $= -4$.

467. $S_n = \left(\dfrac{n}{2}\right)[2a_1 + (n-1)d]$, so $S_{21} = \dfrac{21}{2}[2(1) + (20)(4)]$ (since $a_2 - a_1 = d = 4$) $= \dfrac{21}{2} \times (82) = 861$.

468. $S_8 = \dfrac{a_1(1-r^8)}{1-r} = \dfrac{1\left[1-\left(\frac{1}{3}\right)^8\right]}{1-\frac{1}{3}} = \dfrac{1-\frac{1}{6561}}{\frac{2}{3}} = \dfrac{6560}{6561} \cdot \dfrac{3}{2} = \dfrac{19,680}{13,122}$.

469. These terms are each $\dfrac{1}{2}$ of the previous term, beginning with 2: $\sum\limits_{i=1}^{\infty} 2\left(\dfrac{1}{2}\right)^{i-1}$. When $i = 1$, this is $\left(\dfrac{1}{2}\right)^0 = 1$; when $i = 2$, this is $\dfrac{1}{2}$; etc.

470. $S = \dfrac{a}{1-r}$ where $|r| < 1$. In this case $r = \dfrac{1}{2}$ and $a = 1$, so $S = \dfrac{1}{1-\frac{1}{2}} = \dfrac{1}{\frac{1}{2}} = 2$.

471. $r = \dfrac{1}{9}$, so $|r| < 1$. Thus, $S = \dfrac{a}{1-r} = \dfrac{1}{1-\frac{1}{9}} = \dfrac{1}{\frac{8}{9}} = \dfrac{9}{8}$.

472. Here, $r = 2$; thus $|r| \geq 1$, and S does not exist.

473. $r = -\dfrac{1}{2}$, so $|r| < 1$. $s = \dfrac{a}{1-r} = \dfrac{3}{1-(-\frac{1}{2})} = \dfrac{3}{\frac{3}{2}} = 2$.

474. Remember that $(a+b)^n = \sum\limits_{k=0}^{n} \binom{n}{k} a^{n-k} b^k, n \geq 1$. In this case, $(x+y)^2 = \sum\limits_{k=0}^{2} \binom{2}{k} x^{2-k} y^k = \binom{2}{0} x^2 + \binom{2}{1} xy + \binom{2}{2} y^2$ ($k = 0$, 1, and 2, respectively) $= x^2 + 2xy + y^2$. All problems in which we use the binomial theorem to expand an exponential form of a binomial are done in this way.

475. $(2a+3)^5 = \binom{5}{0}(2a)^5 + \binom{5}{1}(2a)^4(3) + \binom{5}{2}(2a)^3 3^2 + \binom{5}{3}(2a)^2 3^3 + \binom{5}{4}(2a)3^4 + \binom{5}{5}3^5 = (2a)^5 + 5(3)(2a)^4 + 10(2a)^3(9) + 10(2a)^2(27) + 5(2a)81 + 243 = 32a^5 + 30a^4 + 180a^3 + 540a^2 + 810a + 243$.

476. $(2x-4)^3 = \binom{3}{0}(2x)^3 + \binom{3}{1}(2x)^2(-4) + \binom{3}{2}(2x)(-4)^2 + \binom{3}{3}(-4)^3 = 8x^3 + (3)(4x^2)(-4) + 3(2x)(16) - 64 = 8x^3 - 48x^2 + 96x - 64$.

477. **(a)** $P(n, r) = n(n-1)(n-2)\cdots(n-r+1)$. Thus, $P(7, 1) = 7(6)(5)\cdots(7-1+1) = 7$.

(b) $P(n, r) = \dfrac{n!}{(n-r)!} = \dfrac{7!}{(7-1)!} = \dfrac{7!}{6!} = 7$.

478. $P(4, 1) + P(4, 2) + P(4, 3) + P(4, 4) = \dfrac{4!}{3!} + \dfrac{4!}{2!} + \dfrac{4!}{1!} + \dfrac{4!}{0!} = 4 + 12 + 24 + 24 = 64$.

479. $C(n, r) = \dfrac{n!}{r!(n-r)!}$. Thus, $C(31, 2) = \dfrac{31!}{2! \, 29!} = \dfrac{31 \cdot 30}{2!} = 465$.

480. **(A)** There are 5 choices for each digit: $5 \cdot 5 \cdot 5 = 125$.

(B) There are 5 choices for the first digit, 4 for the second, and 3 for the third: $5 \cdot 4 \cdot 3 = 60$.

481. $40 \cdot 30 \cdot 25 = 30{,}000$.

482. Each boy has 4 choices. $4^5 = 4 \cdot 4 \cdot 4 \cdot 4 \cdot 4 = 1024$.

483. The units digit must be 2, 4, or 6. The other digits are any one of 1, 2, 3, 4, 5, 6, 7 with no repetitions: $3 \cdot 6 \cdot 5 = 90$.

484. $7! = 5040$.

485. $\dfrac{7!}{4!} = 7 \cdot 6 \cdot 5 = 210$ (7 objects, 4 objects are alike).

486. There are 10 spots. The first is filled by a boy or girl. There are 10 possibilities. Then there are 5 boys or girls left, and 4 left of the other gender, leaving 9 altogether: $10 \cdot 5 \cdot 4 \cdot 4 \cdot 3 \cdot 3 \cdot 2 \cdot 2 \cdot 1 \cdot 1 = 28,800$.

487. $\dfrac{28,800}{10} = 2880$. In a circle, there is no first or last object.

488. $P(n,\, n) = \dfrac{n!}{(n-n!)} = \dfrac{n!}{0!}; P(n,\, n-1) = \dfrac{n!}{[n-(n-1)]!} = \dfrac{n!}{1!} = \dfrac{n!}{0!}$ since $0! = 1!$

489. Here we are looking for the number of combinations of 17 objects taken 4 at a time. The concern here is not the order of selection or the arrangement of the players. Compare this to, for example, question 485 which is a permutation:

$$C(17,\, 4) = \frac{17!}{4!\, 13!} = 2380.$$

490. See Figure A12.1. Point A can be connected to any of the other points.

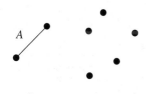

A

Figure A12.1

$$C(9,\, 2) = \frac{9!}{2!\, 7!} = \frac{9 \cdot 8}{2} = 36.$$

491. $C(7,\, 3) \cdot C(4,\, 4) = 35$. [Choose 3 heads, and then the other 4 must all be tails; $C(4,\, 4) = 1$.]

492. $C(9,\, 5) \cdot C(4,\, 4) = \dfrac{9!}{4!\, 5!} \cdot 1 = 126$.

493. **(A)** $C(8,\, 4) = \dfrac{8!}{4!\, 4!} = 70$.

    (B) This is the same as choosing 3 from a group of 7, since we are fixing A as a chosen member: $\dfrac{7!}{3!\, 4!} = 35$.

    (C) Here 2 members are fixed: $C(6,\, 2) = \dfrac{6!}{4!\, 2!} = 15$.

494. $C(21, 5) \cdot C(5, 3) = 20{,}349 \cdot 10 = 203{,}490.$

495. (A) $P(W) = \frac{2}{5}$. Since there are 2 whites (we are looking for the probability of drawing white) and there are 5 balls altogether, $P(E) = \frac{S}{n}$, where E is an event, $S =$ number of ways the event can occur, and $n =$ total number of possibilities.

    (B) $P(B) = \frac{1}{5}$. (There is only 1 black ball; there are 5 balls.)

    (C) $P(G) = \frac{0}{5} = 0$ since there are 0 green balls.

496. (A) $P(H, H, H) = \frac{1}{8}$ since there are $2^3 = 8$ possibilities. Altogether each coin has 2 possibilities, and only one of these is H, H, H.

    (B) There are $2^3 = 8$ possibilities altogether. How many are H, H, T? There are $C(3, 1)$ ways of one coin being T: $C(3, 1) = \dfrac{3!}{1!\,2!} = 3$. Thus, $P(H, H, T) = \frac{3}{8}$. Note: Since $C(3, 1) = C(3, 2)$, we could have used $C(3, 2)$ instead. The number of ways of 1 coin out of 3 being T is the same as 2 coins out of 3 being H.

    (C) $P(T, T, H) = \dfrac{C(3, 2)}{8} = \frac{3}{8}$.

497. (A) $P(\text{all red}) = P(R, R, R)$. There are $C(11, 3)$ possibilities altogether. $C(11, 3) = \dfrac{11!}{3!\,8!} = 165$. Since 6 balls are red, there are $C(6, 3)$ ways to pick R, R, R. $C(6, 3) = 20$. $P(R, R, R) = \frac{20}{165} = \frac{4}{33}$.

    (B) $P(B, B, B) = \dfrac{C(5, 3)}{C(11, 3)} = \dfrac{10}{165} = \dfrac{2}{33}$.

    (C) There are $C(6, 2)$ ways to pick 2 red balls and $C(5, 1)$ ways to pick 1 black ball. Therefore, $P(2R, 1B) = \dfrac{C(6, 2)C(5, 1)}{C(11, 3)} = \dfrac{5}{11}$.

498. (A) $P(R) = \frac{26}{52} = \frac{1}{2}$ since 26 cards are red and 26 are black.

    (B) $P(Q) = \frac{4}{52} = \frac{1}{13}$ since there are 4 queens (1 per suit).

    (C) $P(\text{red } 8) = \frac{2}{52} = \frac{1}{26}$ since only the 8 of hearts and 8 of diamonds are red.

499. (A) $P(\text{both sales}) = \dfrac{2}{3} \cdot \dfrac{1}{2} = \dfrac{2}{6} = \dfrac{1}{3}$.

    (B) $P(\text{no sales}) = \left(1 - \dfrac{2}{3}\right)\left(1 - \dfrac{1}{2}\right) = \dfrac{1}{3} \cdot \dfrac{1}{2} = \dfrac{1}{6}$.

    (C) $P(\geq 1) = P(1 \text{ sale}) + P(2 \text{ sales}) = P(\text{sale, no sale}) + P(\text{no sale, sale}) + P(\text{sale, sale}) = \frac{1}{3} + \frac{1}{3} + \frac{1}{6} = \frac{5}{6}$. [Alternatively, $P(\geq 1) = 1 - P(\text{no sale}) = 1 - \frac{1}{6} = \frac{5}{6}$.] (See question 500 (A))

    (D) $P(1 \text{ sale}) = P(\text{sale, no sale}) + P(\text{no sale, sale}) = \frac{1}{3} + \frac{1}{3} = \frac{2}{3}$.

500. (A) The probability of choosing bag 1 is $\frac{1}{2}$. The probability of choosing a red ball is then $\frac{1}{3}$. Similarly, the probability of choosing bag 2 is $\frac{1}{2}$, but here the probability of choosing a red ball is $\frac{1}{4}$. Therefore, $P(\text{red}) = \frac{1}{2} \cdot \frac{1}{3} + \frac{1}{2} \cdot \frac{1}{4} = \frac{1}{6} + \frac{1}{8} = \frac{7}{24}$.

    (B) $P(\text{white}) = \frac{1}{2} \cdot \frac{2}{3} + \frac{1}{2} \cdot \frac{3}{4} = \frac{17}{24}$. $\left[\text{Alternatively, } P(W) = 1 - P(R) = 1 - \frac{7}{24} = \frac{17}{24}.\right]$